CAMBRIDGE LIBRARY COLLECTION

Books of enduring scholarly value

Botany and Horticulture

Until the nineteenth century, the investigation of natural phenomena, plants and animals was considered either the preserve of elite scholars or a pastime for the leisured upper classes. As increasing academic rigour and systematisation was brought to the study of 'natural history', its subdisciplines were adopted into university curricula, and learned societies (such as the Royal Horticultural Society, founded in 1804) were established to support research in these areas. A related development was strong enthusiasm for exotic garden plants, which resulted in plant collecting expeditions to every corner of the globe, sometimes with tragic consequences. This series includes accounts of some of those expeditions, detailed reference works on the flora of different regions, and practical advice for amateur and professional gardeners.

Planting and Rural Ornament

William Marshall (1745–1818), an experienced farmer and land agent, published this work in 1795, and early in 1796 produced a second edition (reissued here), 'with large additions'. The two-volume work was intended as a practical guide for the owners or managers of large estates on how to establish and maintain timber plantations, both for their financial value and also as important decorative elements in the landscaping of the surroundings of the owner's house. The work covers the practical issues of planting, propagating and transplanting, discusses the choice of trees for different commercial purposes, and the planning and maintenance of hedgerows, as well as ornamental buildings. Volume 1 includes a review of the writings on landscape of such figures as Horace Walpole, (one of whose essays is reproduced), giving insights into the economic as well as the aesthetic aspects of landscape gardening in its golden age.

Cambridge University Press has long been a pioneer in the reissuing of out-of-print titles from its own backlist, producing digital reprints of books that are still sought after by scholars and students but could not be reprinted economically using traditional technology. The Cambridge Library Collection extends this activity to a wider range of books which are still of importance to researchers and professionals, either for the source material they contain, or as landmarks in the history of their academic discipline.

Drawing from the world-renowned collections in the Cambridge University Library and other partner libraries, and guided by the advice of experts in each subject area, Cambridge University Press is using state-of-the-art scanning machines in its own Printing House to capture the content of each book selected for inclusion. The files are processed to give a consistently clear, crisp image, and the books finished to the high quality standard for which the Press is recognised around the world. The latest print-on-demand technology ensures that the books will remain available indefinitely, and that orders for single or multiple copies can quickly be supplied.

The Cambridge Library Collection brings back to life books of enduring scholarly value (including out-of-copyright works originally issued by other publishers) across a wide range of disciplines in the humanities and social sciences and in science and technology.

Planting
and
Rural Ornament

Being a Second Edition, with Large Additions, of Planting and Ornamental Gardening: A Practical Treatise

VOLUME 1

WILLIAM MARSHALL

CAMBRIDGE
UNIVERSITY PRESS

University Printing House, Cambridge, CB2 8BS, United Kingdom

Cambridge University Press is part of the University of Cambridge.
It furthers the University's mission by disseminating knowledge in the pursuit of education, learning and research at the highest international levels of excellence.

www.cambridge.org
Information on this title: www.cambridge.org/9781108075909

This edition first published 1796
This digitally printed version 2015

ISBN 978-1-108-07590-9 Paperback

Selected botanical reference works available in the
CAMBRIDGE LIBRARY COLLECTION

al-Shirazi, Noureddeen Mohammed Abdullah (compiler), translated by Francis Gladwin: *Ulfáz Udwiyeh, or the Materia Medica* (1793) [ISBN 9781108056090]

Arber, Agnes: *Herbals: Their Origin and Evolution* (1938) [ISBN 9781108016711]

Arber, Agnes: *Monocotyledons* (1925) [ISBN 9781108013208]

Arber, Agnes: *The Gramineae* (1934) [ISBN 9781108017312]

Arber, Agnes: *Water Plants* (1920) [ISBN 9781108017329]

Bower, F.O.: *The Ferns (Filicales)* (3 vols., 1923–8) [ISBN 9781108013192]

Candolle, Augustin Pyramus de, and Sprengel, Kurt: *Elements of the Philosophy of Plants* (1821) [ISBN 9781108037464]

Cheeseman, Thomas Frederick: *Manual of the New Zealand Flora* (2 vols., 1906) [ISBN 9781108037525]

Cockayne, Leonard: *The Vegetation of New Zealand* (1928) [ISBN 9781108032384]

Cunningham, Robert O.: *Notes on the Natural History of the Strait of Magellan and West Coast of Patagonia* (1871) [ISBN 9781108041850]

Gwynne-Vaughan, Helen: *Fungi* (1922) [ISBN 9781108013215]

Henslow, John Stevens: *A Catalogue of British Plants Arranged According to the Natural System* (1829) [ISBN 9781108061728]

Henslow, John Stevens: *A Dictionary of Botanical Terms* (1856) [ISBN 9781108001311]

Henslow, John Stevens: *Flora of Suffolk* (1860) [ISBN 9781108055673]

Henslow, John Stevens: *The Principles of Descriptive and Physiological Botany* (1835) [ISBN 9781108001861]

Hogg, Robert: *The British Pomology* (1851) [ISBN 9781108039444]

Hooker, Joseph Dalton, and Thomson, Thomas: *Flora Indica* (1855) [ISBN 9781108037495]

Hooker, Joseph Dalton: *Handbook of the New Zealand Flora* (2 vols., 1864–7) [ISBN 9781108030410]

Hooker, William Jackson: *Icones Plantarum* (10 vols., 1837–54) [ISBN 9781108039314]

Hooker, William Jackson: *Kew Gardens* (1858) [ISBN 9781108065450]

Jussieu, Adrien de, edited by J.H. Wilson: *The Elements of Botany* (1849) [ISBN 9781108037310]

Lindley, John: *Flora Medica* (1838) [ISBN 9781108038454]

Müller, Ferdinand von, edited by William Woolls: *Plants of New South Wales* (1885) [ISBN 9781108021050]

Oliver, Daniel: *First Book of Indian Botany* (1869) [ISBN 9781108055628]

Pearson, H.H.W., edited by A.C. Seward: *Gnetales* (1929) [ISBN 9781108013987]

Perring, Franklyn Hugh et al.: *A Flora of Cambridgeshire* (1964) [ISBN 9781108002400]

Sachs, Julius, edited and translated by Alfred Bennett, assisted by W.T. Thiselton Dyer: *A Text-Book of Botany* (1875) [ISBN 9781108038324]

Seward, A.C.: *Fossil Plants* (4 vols., 1898–1919) [ISBN 9781108015998]

Tansley, A.G.: *Types of British Vegetation* (1911) [ISBN 9781108045063]

Traill, Catherine Parr Strickland, illustrated by Agnes FitzGibbon Chamberlin: *Studies of Plant Life in Canada* (1885) [ISBN 9781108033756]

Tristram, Henry Baker: *The Fauna and Flora of Palestine* (1884) [ISBN 9781108042048]

Vogel, Theodore, edited by William Jackson Hooker: *Niger Flora* (1849) [ISBN 9781108030380]

West, G.S.: *Algae* (1916) [ISBN 9781108013222]

Woods, Joseph: *The Tourist's Flora* (1850) [ISBN 9781108062466]

For a complete list of titles in the Cambridge Library Collection please visit:
www.cambridge.org/features/CambridgeLibraryCollection/books.htm

PLANTING

AND

RURAL ORNAMENT.

VOLUME THE FIRST.

PLANTING

AND

RURAL ORNAMENT.

BEING

A SECOND EDITION,

WITH

LARGE ADDITIONS,

OF

PLANTING AND ORNAMENTAL GARDENING,

A PRACTICAL TREATISE.

IN TWO VOLUMES.

VOLUME THE FIRST.

LONDON:

Printed for G. Nicol, Bookseller to his Majesty, Pall-Mall; G. G. and J. Robinson, in Paternoster Row; and J. Debrett, Piccadilly.

M,DCC,XCVI.

CONTENTS

OF THE

FIRST VOLUME.

SUBJECT THE FIRST.

PLANTING.

DIVISION THE FIRST.

MANUAL OPERATIONS.

DIVISION THE SECOND.

CHOICE OF TIMBER TREES.

DIVISION THE THIRD.

HEDGES AND HEDGEROW TIMBER.

SECT.

DIVISION

DIVISION THE FOURTH.

WOODLANDS OR USEFUL PLANTATIONS.

The

On

Cultivating

SUBJECT THE SECOND.

RURAL ORNAMENT.

DIVISION THE FIRST.

HISTORY OF THE RURAL ART.

DIVISION THE SECOND.

PRINCIPLES OF THE RURAL ART.

Definition

Principles

Seats

DIVISION THE THIRD.

APPLICATION OF THE RURAL ART

DIVISION

DIVISION THE FOURTH.

REMARKS ON PLACES.

DIVISION THE FIFTH.

MINUTES IN PRACTICE.

ADVER-

ADVERTISEMENT.

THE Intention of this Publication is to bring into one point of view, and arrange in a compendious form, the Art of Planting and Laying-out Plantations: an art which, though in itself a unity, has hitherto been treated of as two distinct subjects. Books on Planting we have many; and those on Ornamental Gardening are not less numerous; but a Practical Treatise, comprehending the entire subject of conducting Rural Improvements, upon the principles of modern taste, has not hitherto appeared in public. This circumstance, however, is the less to be wondered at, as the man of business and the man of taste are rarely united in the same person. There are many Nurserymen who are intimately acquainted with the various methods of propagating trees and shrubs; and many

Gentlemen whoſe natural taſte, reading, and obſervation enable them to form juſt ideas of rural embelliſhment; but where ſhall we find the Nurſeryman who is capable of ſtriking out the great deſign, or the Gentleman equal to the management of every tree and ſhrub he may wiſh to aſſemble in his collection? To proceed one ſtep farther, where is the Gentleman, or the Nurſeryman, who is ſufficiently converſant in the training of Woodlands, Hedges, and the more uſeful Plantations? In fine, where ſhall we look for the man who in the ſame perſon unites the Nurſeryman, the Woodman, the Ornamentaliſt, and the Author? We know no ſuch man: the reader, therefore, muſt not be diſappointed when he finds, that, in treating of exotic trees and ſhrubs, the works of preceding writers have been made uſe of.

COOK is our firſt writer on Planting; nevertheleſs EVELYN has been ſtyled the Father of Planting in England. It is probable that, in the early part of life, EVELYN was a practical planter, upon his eſtate at Wotton in Surrey; but his book was written in the wane

wane of life, at Greenwich, during a long and painful fit of the gout. His *Sylva* contains many practical rules, valuable, no doubt, in his day, but now superseded by modern practice; and may be said to lie buried in a farrago of traditional tales, and learned digressions, suited to the age he lived in *. MILLER at length arose among a group of minor planters; and after him the indefatigable HANBURY, whose immense labours are in a manner lost to the Public.

COOK and EVELYN treated professedly of FOREST TREES, MILLER and HANBURY include ORNAMENTALS; but their works, which are voluminous and expensive, also include kitchen gardening, flower gardening, the management of greenhouses, stoves, &c. &c. the propagation of trees and shrubs, adapted to the open air of this climate, forming only a small portion of their respective publications.

MILLER

* The first Edition was printed in 1664.

MILLER and HANBURY, however, are the only writers who could afford us the required aſſiſtance; and we were led to a choice of the latter, as our chief authority, by three principal motives:—HANBURY wrote ſince MILLER, and, having made ample uſe of Mr. M.'s book, his work contains, in effect, the experience of both writers: MILLER is in the hands of moſt Gentlemen; HANBURY is known to few; his book, either through a want of method, a want of language, or through an ill judged plan of publiſhing on his own account, has never *ſold:* and laſtly, MILLER's botanical arrangement is become obſolete; HANBURY's is agreeable to the Linnean ſyſtem.

SINCE MR. HANBURY's death, the Public have been favored with a new and ſumptuous edition of EVELYN's *Sylva*; with notes by Dr. HUNTER of York, conſiſting of botanical deſcriptions, and the modern propagation of ſuch trees as EVELYN has treated of. Theſe notes, however, contain little new information; the deſcriptions being principally copied from MILLER, and the practical directions from HANBURY.

LEST unacknowledged aſſiſtance, or aſſiſtance acknowledged indirectly, ſhould be laid to our charge, it is thought proper to particularize, in this place, the ſeveral parts of this publication, which are *written*, from thoſe which are *copied*.

THE INTRODUCTORY DISCOURSES, containing the MANUAL OPERATIONS of Planting, and the OUTLINE of the LINNEAN SYSTEM, are, as rudiments, entirely new; excepting the quotations from Linneus's work, which quotations are extracted from the Lichfield Tranſlation of The *Syſtema Vegetabilium* of that great man.

THE ALPHABET OF PLANTS, ſo far as it relates to TIMBER TREES, and other NATIVE PLANTS, as well as to ſome of the more USEFUL EXOTICS, is either wholly our own, or contains ſuch additions, as have reſulted from our own obſervation and experience: ſo far as it relates to ORNAMENTAL EXOTICS, it is entirely HANBURY'S; excepting the quotations which are marked, and excepting the GENERAL ARRANGEMENT, which

which is entirely new. HANBURY has not leſs than ſix diſtinct claſſes for the plants here treated of, namely, deciduous Foreſt Trees, Aquatics, evergreen Foreſt Trees, deciduous Trees proper for ornament and ſhade, evergreen Trees proper for ornament and ſhade, and hardy climbing Plants. The firſt three claſſes are without any ſubordinate arrangement; in the laſt three the plants are arranged alphabetically, agreeably to their genera. This want of ſimplicity in the arrangement renders the work extremely heavy, and irkſome to refer to; and is productive of much unneceſſary repetition, or of tireſome references, from one part of his unwieldy work to another. His botanical ſynonyms we have wholly thrown aſide, as being burthenſome, yet uninſtructive; and, in their place, we have annexed to each Species the trivial or ſpecific name of LINNEUS; which, in one word, identifies the plant, with a greater degree of certainty, than a volume of Synonyma. Other retrenchments, and a multiplicity of corrections, have taken place: however, where practical knowledge appears to ariſe incidentally out of our author's own experience, we have

have cautiously given it in his own words: likewise, where interesting information lies entangled in a singularity of manner, from which it could not well be extricated, we have marked the passages containing it, as literal quotations;—to distinguish them from others, which, having been written in a manner more properly didactic, or brought to that form by retrenchment or correction, we consider as being more fully intitled to the places we have assigned them.

THE articles TIMBER, HEDGEROWS, and WOODLANDS, are altogether new *, being drawn from a considerable share of experience, and an extended observation.

THE Sections on RURAL ORNAMENT are likewise new, if anything new can be offered on a subject, upon which so much has been already written. Taste, however, is a subject upon which all men will think and write differently, even though their sources of information

* Excepting such extracts and quotations as are marked, and have their respective authorities subjoined.

mation may have been the ſame. WHEATLEY, MASON, and NATURE, with ſome EXPERIENCE, and much OBSERVATION, are the principal ſources from which this part of our work was drawn: if we add that it was planned, and in part written, among the magnificent ſcenes of nature, in Monmouthſhire, Herefordſhire, and Glouceſterſhire, where the rich and the romantic are happily blended, in a manner unparalleled in any other part of the Iſland, we flatter ourſelves no one will be diſſatisfied with the *origin:* of the *production*, let the Publick ſpeak.

To this SECOND IMPRESSION, we have been enabled to make conſiderable ADDITIONS; particularly to the Subject RURAL ORNAMENT. The REMARKS on ORNAMENTED PLACES, as well as the MINUTES on our own PRACTICE, which are now firſt printed, are tranſcribed from the rough *memoranda*,

moranda, that were written at the times of obſervation, or as the incidents and reflections occurred.

ON the ſubject of PLANTING, too, will be found ſome additional information; more eſpecially in the Sections WOODS, and TIMBER GROVES.

IT may alſo be right to mention, here, that we have omitted to inſert, in this Edition, Mr. FARQUHARSON's Paper on the propagation of the *Scotch Fir*; a tree which, now, when the ſuperior merits of the *Larch* are aſcertained, can ſeldom be *planted* with propriety.

WE have likewiſe thought it right to omit ſome remarks on the SALE AND FELLING OF TIMBER; a ſubject which does not properly belong to *Planting*, and is much leſs compatible with *Rural Ornament*. We therefore confine this Work to the PRODUCTION OF WOODLANDS, whether uſeful or ornamental, and

and refer the Reader, for their GENERAL MANAGEMENT, a ſubject in itſelf of great extent and importance, to the different Works which we have publiſhed on RURAL ECONOMY: a liſt of which will appear at the cloſe of theſe Volumes.

LONDON, *December* 1795.

PLANTING

AND

RURAL ORNAMENT.

General View of the Subjects.

THE earth produces an almoſt infinite variety of Plants, poſſeſſing various properties, and different degrees of ſtrength and ſtature. In the vegetable, as in the animal world, the ſtronger ſubdue the weaker: the herbaceous tribes bow to the ſhrub, and this to the more robuſt foreſt tree; and, in an unpeopled country, a ſtate of woodineſs prevails. The interior parts of America are at this day a foreſt: the Continent of Europe, too, has ſtill its foreſt; and England once was famous for her's.

As inhabitants increaſe, woodineſs gives way to huſbandry and the arts; not merely as an incumbrance, but as affording uſeful materials. Population ſtill increaſing, the *foreſt* breaks into *woods*. Commerce and luxury advancing, the canoe becomes a ſhip, and the hut a manſion: at length even the woods dwindle away, and *plantations*, or an *import of foreign timber*, become neceſſary to ſupply the want.

ENGLAND has experienced, more or leſs, every ſtage of this decline. Its preſent ſtate, in reſpect to timber, we conceive to be this: A few broken foreſts, and many extenſive woodlands, ſtill remaining; a great number of plantations of different growths, and a vaſt ſupply of foreign timber of various kinds. Indeed, we are of opinion, that had it not been for this foreign ſupply, ſcarcely a timber tree, at this day, would have been left ſtanding upon the iſland.

OUR exiſtence, as a nation, depends upon a full and *certain* ſupply of ſhipping; and this, we may venture to ſay, upon an *internal* ſupply of ſhip timber. That there is no want of oak timber, at preſent, in this iſland, is, we believe, a fact; but that the article of *ſhip* timber is growing ſcarce, as we ſhall explain more fully in its proper place, is, we believe, alſo a fact which cannot be controverted. This is an important matter, which demands the firſt attention of Government, and is is not unworthy the notice of every landed individual.

MANKIND, however, do not view the face of nature in the light of ſelfpreſervation only ; the great Author of creation has wonderfully adapted our ſenſes to the enjoyment of its delights ; the eye is gratified by tints of verdure, and the ear by the muſic of the woods and the mellowneſs of echo—and both by the voice and majeſty of a foreſt, rouſed by the breath of Nature. Our plan therefore, has two objects, UTILITY and ORNAMENT ; they are nearly allied, however, as exerciſe and recreation, or as the uſe and the ornament of dreſs,

NEVERTHELESS, to treat of them with greater advantage, it will be proper to conſider them ſeparately, as two diſtinct ſubjects.

SUBJECT THE FIRST.

PLANTING.

DIVISION THE FIRST.

MANUAL OPERATIONS.

INTRODUCTORY REMARKS.

BEFORE we attempt to give directions for cultivating WOODLANDS, or raiſing ORNAMENTAL PLANTATIONS, it will be proper to give a comprehenſive view of the MANUAL OPERATIONS incident to

PROPAGATING, PLANTING and
TRAINING TRANSPLANTING
Trees and Shrubs in general.

BUT before the young planter put his foot upon the ſpade, we beg leave to caution him, in the ſtrongeſt terms, againſt a WANT OF SPIRIT in Planting. A ſlovenly planter ranks among the moſt extravagant order of ſlovens: the labour, the plants, and the ground are thrown away; beſides the conſequent diſgrace, not only to the individual, but to the profeſſion. Anxious and intereſted as we are in the cauſe of planting, we would rather want pupils, than have them paſs through our hands *unfiniſhed:* we therefore reject all ſuch as have not induſtry, ſpirit, and perſeverance, to go through with what they undertake; and we recommend to ſuch as are poſſeſſed of theſe valuable qualifications, *to begin upon a ſmall ſcale,* and to let their ſeminaries, their nurſeries, and their plantations, increaſe with their experience.

WHILE, however, we caution our readers againſt entering, immaturely, upon the buſineſs of planting, we cannot refrain from mentioning the PLEASURES which reſult from it. How rational, and to a contemplative mind how delightful, to obſerve the operations of Nature;—to trace her in every ſtage, from the ſeed to the perfected plant; and, from beneath the leaf ſtalk of this, through the flower bud, the flower, and the ſeed veſſel, to the ſeed again! Man muſt be employed; and how more agreeably than in converſing with Nature, and in ſeeing the works of his own hands, aſſiſted by her, riſing into perfection.

NOR

NOR do we mean to hold out pleaſure, alone, as an inducement to planting;—its PROFITS are great, when properly executed; and this idea adds ſolidity to the enjoyment. Pleaſure alone may ſatiate; but profit and pleaſure, united, ſeldom fail of producing a laſting gratification.

THERE is another incitement to planting, which alone has been generally held out as a ſufficient inducement. We are ſorry to confeſs, however, that we know too much of mankind to believe that PATRIOTISM, unaided by perſonal intereſt, will ever produce a ſupply of ſhip timber to this or any other nation. Far be it from us, however faſhionable it may be, to ſpeak irreverently of patriotiſm; we conſider it as the nobleſt attribute of the human mind. Young men, to whom we more particularly addreſs ourſelves, are ſeldom without ſome ſhare of it; and we flatter ourſelves that this virtuous principle, aſſiſted by the pleaſure, the profit, and the POPULARITY which attends planting, will induce the young men of the preſent age to ſtudy and practiſe it; not more for themſelves, than for future generations.

SECTION THE FIRST.

PROPAGATING TREES AND SHRUBS.

TREES and SHRUBS are propagated

From SEEDS,	By LAYERING,
—— SUCKERS,	— BUDDING,
—— CUTTINGS,	— GRAFTING.

I. PROPAGATING FROM SEED.—There are four ways of raiſing, from ſeed, the trees and ſhrubs adapted to our purpoſe:

In Beds of natural Soil,
In Beds of Compoſt,
In Pots,—and ſome few
In Stoves, or under Glaſſes.

IT will be expected, perhaps, before we begin to treat of the different methods of ſowing, that we give ſome directions for GATHERING and preſerving ſeeds. Little, however, can be ſaid upon the ſubject under this general head; different ſpecies requiring a difference in management. We may, nevertheleſs, venture to ſay, that all ſeeds ought

ought to be fully matured upon their native plants; and we may further add, that ſuch as drop ſpontaneouſly from the ſeed veſſel, or are ſhed by a moderate wind, or other gentle agitation, are preferable to thoſe which are torn from the tree, immaturely. The ſeeds of ſcarce, or valuable plants may be gathered thus: As ſoon as they begin to fall, voluntarily, ſpread a cloth under the plant, and agitate it moderately, until all that are ripe have fallen;—and repeat this, whenever a ſecond, and a third, ſpontaneous fall takes place.

THE art of PRESERVING ſeeds reſts chiefly upon that of *curing* them, immediately after gathering. If graſs were put into the ſtack immediately, after mowing, or corn threſhed out, at harveſt, and laid in heaps, it would preſently heat, and be entirely ſpoiled. So it is with the ſeeds of trees and ſhrubs: therefore, they ought, as ſoon as they are gathered, to be ſpread thin, in an airy place, and be turned, as often as a cloſe attention ſees neceſſary. When the ſuperfluous moiſture has evaporated, they may be collected into bulk; remembering, however, to run them frequently down a ſkreen, or ſhake them in a ſieve, that their brightneſs and ſweetneſs may be preſerved. Some of the larger ſeeds, acorns eſpecially, are difficult to cure, and require a very ſtrict attention.—It muſt alſo be remembered, that mice, and other vermin, are dangerous

enemies

enemies to ſeeds. Thoſe which are particularly valuable, may be hung up, in bags, to the ceiling of a dry room.

IN PROCURING SEEDS from the SHOPS, or from ABROAD, ſome caution is neceſſary. A ſeedſman, who has a character to loſe, and a correſpondent, who is himſelf a judge of the quality of ſeeds, are the beſt general guards againſt impoſition and diſappointment.

THERE are ſeveral ways of TRYING THE QUALITY of ſeeds. The heavier kinds may be proved in water; ſuch as ſwim are at leaſt doubtful. The lighter ſorts may be tried by biting them: if they break abruptly between the teeth, they are generally good; but if they be tough and leathery, they are moſtly the contrary. If when cruſhed, or ſeparated by a knife or ſciſſars, they appear firm, white, and farinaceous, they may generally be eſteemed good; but if, on the contrary, they be ſpungy and diſcoloured, they are generally of a bad quality. But the moſt certain mode of trial, and that which in caſes of ſuſpicion ought never to be omitted, is to *force* a few of them, in a garden pot, placed in an artificial heat, or other warm ſituation. Put in ſome certain number, taken promiſcuouſly from the parcel, and, from the proportional number that vegetate, a tolerably juſt idea may

may be formed of the quality of the whole. Without this precaution, a ſeaſon may be loſt, and the uſe of the land, together with the labour, be thrown away.

ALL the natives, and many exotics, may be raiſed in BEDS OF NATURAL MOLD. The *ſoil* ſhould be rich, and ſufficiently deep to admit of being trenched, or double dug, two ſpits deep. If it will not bear one ſpit and a half, namely about fourteen inches, it is improper for ſeed beds, and ſhould either be wholly rejected, or (if the ſubſtratum is not of too hungry and poiſonous a nature) be trenched, a ſpit and a half deep, and the crude mold meliorated, by manure, and repeated diggings. Autumn is the beſt time to bring up the ſubſtratum, letting it lie in rough trenches all winter, to take the froſt. In the ſpring, put on a quantity of dung, in proportion to the poverty of the ſoil; turning it in, ſuperficially, and mixing it well with the ſoil to be improved. Repeat this ſingle digging, through the ſummer, as often as convenient, or as often as the weeds, which never fail to riſe, in great abundance, from a ſubſtratum expoſed to the ſun and air, require it. In autumn, turn up the ſoil from the bottom, and mix the whole well together. The longer the ſoil and ſubſtratum lie in the ſtate of inverſion, the better tempered the freſh mold will become, and the mellower

mellower will be the old cultivated ſoil. In a manner ſimilar to this, all ſoils, which are not naturally rich, ought to be treated. No department of planting calls more loudly for a ſpirited management than the ſeminary; which, if not rich and deep by nature, ought to be made ſo by art, at almoſt any expence.

In large undertakings, a ſeparate *ſeminary* may be neceſſary; but, in general, a portion of the kitchen garden is better adapted to the purpoſe. There are, indeed, two very great advantages, in mixing the ſeminary with the kitchen garden: the ſeed beds are always under the eye, and are more likely to be defended from weeds and vermin, there, than in a detached ſeminary, viſited only now and then; and, when the ground has borne a crop of ſeedling plants, it may be applied to the purpoſe of culinary herbs; while that which has been long under crops of theſe, may be changed to nurſery beds. In whatever ſituation it is placed, it muſt be carefully fenced againſt hares and rabbits, or the labour of a whole ſeaſon may be cut off in a few nights: in this light, alſo, the kitchen garden has a preference.

It would be idle to give particular directions for laying out a ſeminary, or to ſay, under this general head, where this or that ſeed ſhould be ſown. Suf-

Suffice it, therefore, to mention, here, that *ſeed beds* are generally made from four to four feet and a half wide, with intervals of one foot and a half to two feet. Theſe dimenſions render them convenient to be weeded, without the plants being trodden or kneeled upon.

THE *methods of ſowing* are various; as DIBBLING, DRILLING, and BROADCAST; which laſt is the moſt prevailing method. Seeds ſown in the promiſcuous broadcaſt manner, are covered either with the rake, or with the ſpade (or ſieve). COVERING WITH THE SPADE (or ſieve) is the common practice, and is thus performed: The ſurface being made light and fine, by a recent digging and raking, and the beds formed (operations which every gardener and gardener's man are acquainted with), a thin coat of mold is raked off the beds, into the intervals, in proportion to the depth the ſeeds require to be buried, and according to the nature of the ſoil, taken jointly. In a light ſandy ſoil, the ſeeds require to be buried deeper than they do in a ſtrong loam; and while an acorn may be covered from one to three inches deep, the ſeeds of the Larch will not bear more than from a quarter to three fourths of an inch. The new ſurface being rendered perfectly fine and level, the ſeeds are ſown, and, in ſome caſes, preſſed gently into the mold, by patting it with the back of the ſpade. The

The earth, which was raked off into the interval (or taken off with a ſpade and placed in little hillocks in it) is now returned; either by caſting it on with the ſpade, with a kind of ſleight which nothing but practice can give, or by ſifting it on, through a ſieve (an operation more eaſy to the inexpert, and in many caſes preferable) as even and regularly as poſſible. The intervals cleared, the beds neated up, and, if the ſoil be light, or the ſeed requires it, their ſurfaces patted with the back of the ſpade, ſo as to give them a kind of poliſhed firmneſs, the buſineſs is finiſhed. DRILLING is performed two ways: By drawing open drills, with hoes, in the common manner; or by taking off the ſurface of the beds, drawing lines upon the new ſurface, laying or ſcattering the ſeeds along theſe lines, and covering them with the ſpade or ſieve, as above directed for broadcaſt ſowing. DIBBLING requires no deſcription.

THE next buſineſs of the ſeminary is to *defend* the ſeed and ſeedlings from *birds*, *vermin*, the *weather*, and *weeds*. Nets are the beſt guard againſt birds, and traps againſt vermin. As a defence againſt the ſcorching heat of the ſun, the beds ſhould be hooped, and mats occaſionally ſpread over them, in the manner of a tilt or awning; but, when the ſcorching abates, the mats ſhould be taken off, to give the plants the benefits of the

at-

atmoſphere; and, in dry weather, the beds ſhould be kept conſtantly watered. The awnings are equally ſafeguards againſt ſpring froſts, than which nothing is more injurious to ſeedling plants. In reſpect to WEEDS, there is a general rule, which ought not to be departed from; that is, not to ſuffer them to get too ſtrong, before they be drawn; for, if they be permitted to form large roots, they not only encumber and rob the ground, but, in drawing them, many ſeeds, or tender ſeedlings, will be drawn out with them. To prevent the young plants from being DRAWN OUT OF THE GROUND BY WINTER FROSTS, which they are very liable to, eſpecially by a continuance of froſt and thaw, alternately, coal aſhes may be ſifted over them. If this evil has already taken place, and the roots appear expoſed above ground, ſome fine mold ſhould firſt be ſifted on, to cover the roots, and then the aſhes ſifted over the mold. If the plants be BEATEN OUT OF THE GROUND BY HEAVY RAINS, the remedy is ſimilar.

THE length of time between the ſowing of the ſeed and the appearance of the plant, is very uncertain: much depends upon the ſeaſon, and ſtill more upon the nature of the plant itſelf. Some ſeeds lie in the ground a whole year before they vegetate, and ſome two or three years,—as will be mentioned under their reſpective Species. During

ing this time the beds ſhould be kept free from weeds and moſs; and, in caſe of a long continuance of dry weather, ſhould be well watered. After very heavy rains, which are liable to run the ſurface to a batter, and waſh away part of the ſoil, it is well to rake the beds ſlightly, and ſift over them a little freſh mold: this prevents the ſurface from baking, and at once gives a ſupply of air, and nouriſhment, to the embryo plants.

BEDS OF COMPOST are made by mixing drift ſand, or other materials, with the natural ſoil of the ſeminary; or with virgin mold, taken from a rich meadow, or old paſture ground. But the particular ingredients of a compoſt depend upon the nature of the plant to be raiſed; and the reader is referred to the reſpective Species, in the ALPHABET OF PLANTS, for further information on this head.

THE mode of raiſing plants, in POTS and BOXES, alſo depends greatly upon the particular plant to be raiſed. The chief intent of this method is to guard the embryo and ſeedling plants from the extremes of heat and cold. The pots are filled with compoſt, ſuited to the plant. For examples, ſee the articles ANNONA, ARALIA, AZALIA, MELIA, PISTACIA, &c. &c.

II. PRO

II. PROPAGATING FROM CUTTINGS.

It is not from ſeeds, alone, that plants may be increaſed; ſo great a ſimplicity prevails in the ſyſtem of vegetation, that numerous tribes may be propagated from twigs or truncheons, cut out of the woody parts of the plants themſelves, and ſtuck naked into the ground, without either root or branch upon them: the part placed in contact with the ſoil ſends forth roots, while that which is expoſed to the open air, throws out branches!

BUT altho' moſt of the aquatics, and many other genera of trees and ſhrubs, may be raiſed from CUTTINGS, planted in common earth and in the open air, there are others which require more care and greater helps. Some require a warm, others a cool border: ſome muſt be rooted in pots, others in ſtoves, or in the greenhouſe. Again, ſome ſhould be taken from the older branches, others from younger ſhoots: ſome require to be planted in autumn, others in the ſpring. Theſe and other peculiarities of treatment will be ſpecified, when we come to treat, ſeparately, of each individual.

III. PROPAGATING FROM SUCKERS.

There is a great ſimilarity between the branches and the roots of plants. If the fibres of ſome ſpecies become expoſed to the air, they quit their function of ſupplying the parent plant with nouriſhment, and, taking upon them the nature of ſeedlings, put forth leaves and branches. Theſe rootling plants are called SUCKERS; and if they be ſlipt off from the parent root, and planted in a ſoil and ſituation ſuited to their reſpective natures, they will grow up, in the manner of ſeedling plants.

VARIOUS opinions are held, reſpecting the propriety of raiſing trees and ſhrubs from ſuckers: EVELYN and MILLER are againſt the practice; ſaying, that plants raiſed from ſuckers are more apt to ſend up ſuckers (which are troubleſome intruders, eſpecially in ornamental grounds) than thoſe of the ſame ſpecies which have been raiſed from ſeeds. HANBURY, however, is of a contrary opinion: he ſays, "What might incline people to this notion was, that they have obſerved trees raiſed from ſeeds very long before they produced ſuckers: but they ſhould conſider, that no tree or plant will produce ſuckers, till it is of a ſuitable ſize or ſtrength for the purpoſe, any more than animals can produce young before they are of proper age; and let them plant a ſeedling that is grown ſtrong, a layer of the ſame ſtrength, and one which has been raiſed from a ſucker, exactly of the ſame ſize, and with the ſame number of fibres to the root,

root, and they will find that the ſeedling, or the layer, will not be behind-hand with the other in producing ſuckers, if they have all a like ſoil and ſituation; for it is peculiar to them to ſport under the ſoil, in this manner; and Nature will ever act agreeably to herſelf, if not ſtopped in her progreſs by art." Nevertheleſs, in ſpeaking of particular plants, we find him holding forth a different language.

IV. PROPAGATING BY LAYERING.—As the roots of ſome plants, when expoſed to the air, ſend forth ſhoots and branches, ſo the branches of others, when placed in contact with the earth, ſend out fibres and roots, which being ſevered from the parent plant, a ſeparate tree is produced.

LAYERING being an operation by which a great majority of trees and ſhrubs may be propagated, and by which the many beautiful variegations are principally preſerved, we ſhall here give ſome general directions for performing it; reſerving, however, the minutiæ, peculiar to each ſpecies, until we come to treat of the individual ſpecies, ſeparately.

LAYERS are bent, either from the *ſtools* of trees and ſhrubs, headed down to a few inches above the ſurface of the ground, or from *boughs*, plaſhed ſo as to bend their tops to the ground; or from *trees* brought into a ſtooping poſture, by excavating the ſoil on one ſide of them, until their heads are lowered into a ſimilar ſituation.

STOOLS afford the ſimpleſt, and are the moſt common, ſupply of layers. Where a great number of layers are wanted, plants ſhould be raiſed for the purpoſe, and planted in ſome well fenced ground, or in ſome vacant part of the ſeminary, or nurſery; and, when of a proper age and ſize, be headed down, to the height of about eight inches, for ſtools. In many caſes, trees ſtanding in grounds, or woods, may be cut down, and give a ſufficient ſupply. In whatever ſituation they are, the earth round them muſt be doubly dug, as deep as the ſoil will allow, and be treated in a manner ſimilar to that of a ſeed bed.

THE METHOD OF LAYERING is this: Dig a ſhallow trench round the ſtool (of a depth ſuitable to the ſize and nature of the plant, as from four to eight inches), and having pitched upon the ſhoots to be layered, bend them to the bottom of the trench (either with or without plaſhing, as may be found moſt convenient), and there *peg* them faſt; or, putting ſome mold upon them, *tread* them hard enough to prevent their ſpringing up again;

again;—fill in the mold;—place the top of the layer in an upright posture, treading the mold hard behind it; and cut it carefully off, above the second or third eye.

IN this simple way a numerous tribe of plants may be layered: there are many, however, which require a more complex treatment. Some will succeed by having a *chip* taken off the under side of the lower bend of the layer, which gives the fibres an opportunity of breaking out, with greater freedom: others, by having a *cleft* made, in that part, by thrusting an awl or bodkin through it, keeping the cleft open, by a chip or wooden pin; or by making a longitudinal *slit in the bark* only: others succeed better, by *twisting* the part: and others, again, by *pricking* it, and binding a *wire* round it. But when SIMPLE LAYERING will not succeed, the most prevailing, and in general the most certain, method is that of TONGUE LAYERING; which is thus performed: The excavation being made, and the layer chosen and trimmed, ascertain where the lower bend of it will fall, by taking it in the left hand and bending it down to the bottom of the trench; then placing the thumb of the right hand firmly against the part opposite which the tongue falls, insert the edge of the knife, as with an intent to cut the layer off short, in that place; but having cut about half way thro' it, turn the edge of the knife abruptly upwards,

 drawing

drawing it along the pith, half an inch, or an inch, according to the ſize of the layer. The whole ſtool being treated in this manner, proceed to peg the layers cloſe to the bottom of the trench, bedding the cleft or mouth of each in fine mold, for the fibres to ſtrike into. (If the mold and the ſeaſon be very dry, it may be well to moiſten ſome fine mold with ſoft water, making it into a paſte, and wrap the wounded part in a handful of this prepared earth.) This done, level in the mold, draw the point of the layer upright, and ſhorten it, as above directed; being careful to diſturb the wounded part as little as poſſible. It is a practice with ſome to clear the ſtools, entirely, after layering: we would rather recommend, however, to trim off ſuch ſhoots only as are too old, or are defective, leaving ſuch as are too young, to increaſe in growth; by which means an annual, inſtead of a biennial, ſucceſſion of layers will be had.

THE TIME OF LAYERING is generally autumn; ſpring is favourable to ſome plants, and midſummer to others; but trees and ſhrubs, in general, may be layered at almoſt any time of the year.

THE length of time requiſite for ROOTING a layer depends upon the nature of the plant: twelve months is generally conſidered as a ſufficient time, during which the layers ſhould be kept clear from weeds;

weeds; and, when the rooted plants are taken off, the ſtumps, from which they were ſevered, ſhould be cut off cloſe to the ſtools, in order that they may ſend forth a future ſupply of ſhoots.

V. VI BUDDING—AND GRAFTING are operations more particularly applicable to fruit trees, and belong to the *kitchen gardener* rather than to the *planter*. They are operations difficult to deſcribe upon paper; and are known to every nurſeryman and gardener. The great art in *grafting* lies in uniting the graft cloſely, and firmly, with the ſtock; and in *budding*, not to leave too much wood, nor yet to pare it off too cloſe to the eye.

SECTION THE SECOND.

TRAINING TREES AND SHRUBS.

TREES and SHRUBS may be trained up from the ſeed bed, &c. until they be fit to be planted out to ſtand, either in NURSERIES ſet apart for the purpoſe, or in YOUNG PLANTATIONS; which laſt are frequently the moſt eligible nurſeries, as will

 be

be explained hereafter. A SEPARATE NURSERY however, is nevertheless necessary; and in this place it will be proper to give some general ideas of the soil, situation, and business of a nursery ground

THE SOIL of the nursery, like that of the seminary, should be rich and deep, and like that, also, should be prepared, by double diggings, and suitable meliorations: if not deep and rich by nature, it must be made so by art, or be wholly rejected, as unfit for the purposes of a nursery ground. For, if the roots of the tender plants have not a soil they affect, or a sufficient room to strike in, there will be little hopes of their furnishing themselves with that ample stock of fibres which is necessary to a good plant, and with which to supply them is the principal use of the nursery.

THE SITUATION of the nursery is frequently determined by the soil, and frequently by local conveniencies: the nearer it is to the garden or seminary, the more attendance will probably be given it; but the nearer it lies to the scene of planting, the less carriage will be requisite. In whatever situation the nursery be placed, it must, like the seminary, be effectually fenced against hares and rabbits.

THE

THE BUSINESS OF THE NURSERY consists principally in

PREPARING THE SOIL,	PRUNING THE NURSERY PLANTS,
PRUNING THE SEEDLINGS, &c.	THINNING THEM,
PUTTING THEM IN,	TAKING UP, AND
KEEPING THEM CLEAN,	PACKING FOR CARRIAGE.

THE PREPARATION OF THE SOIL has already been mentioned: too much pains cannot be taken, in this department; it is the foundation, upon which the success of the whole business greatly depends.

IN PRUNING seedlings, layers, and suckers, for the nursery, the ramifications of the roots should not be left too long and sprawling; but, *in this case,* should be trimmed off pretty close, so as to form a snug globular root. By this means, the new fibres will be formed immediately round the root of the plant, and may, of course, be easily removed with it, and without disturbing the earth interwoven among them. The tops should, in most cases, be trimmed quite close up to the leader, or (if aukward or defective) be cut off a little above the root.

IN PUTTING IN seedlings, various methods are practised: by the *dibble*; by the *scoop*; by a single chop

chop with the *spade*, or by two chops, one across the other: by *square holes*, made by four chops of the spade, bringing up the mold with the last; or by *bedding*; a method chiefly made use of for quicksets. If the soil be well prepared, and the plants properly pruned, the chief art, in putting them in, lies in not cramping the fibres of the roots; but, on the contrary, in letting them lie free and easy, among the mold: and the particular mode, or instrument to be made use of, depends much upon the size of the plants to be put in. This also determines, in a great measure, the proper *distance* between the rows and between plant and plant. Strong suckers or layers require larger holes, and a greater distance, than weak seedling plants. The proposed method of cleaning, too, is a guide to the distance: the plow cannot work in so narrow a compass as the spade. The natural tendency of the plant itself must also be considered; so that few general directions can be given, under this head. If we say from six to twenty-four inches in the rows, with intervals from one to four feet wide, we shall comprehend the whole variation of distances.

CLEANING THE NURSERY is a business which must not, of all others, be neglected: all plants are enemies to each other. If grass and weeds are suffered to prey upon the soil, the young plants will

will be deprived of their proper nouriſhment and moiſture: in ſhort, it is neceſſary that the nurſery ſhould be kept equally clean as the ſeminary, and this as clean as the kitchen garden: it would be more pardonable to ſuffer the plants to be ſmothered in the ſeed bed than in the nurſery quarters; for, in that caſe, only a ſmall part of the expence would be thrown away. Nor is merely keeping the weeds under, the only care in a nurſery: the intervals muſt be kept ſtirred, in order to give air and freedom to the fibres. This may be done either with the ſpade, which is called turning in; or, if the intervals be wide enough, and the nurſery extenſive, with the plow, which is attended with much leſs expence.

THE next buſineſs of the nurſery is PRUNING: this is neceſſary, to prevent the plants from crouding each other, and to give them ſtem. Shrubs, which do not require a ſtem, ſhould not be placed in nurſery rows, but in the quincunx manner, that they may have an equal room to ſpread, on every ſide; but foreſt trees, and trees in general, require ſome length of ſtem; and, in giving them this, the leading ſhoot is more particularly to be attended to. If the head be double, one of the ſhoots muſt be taken cloſe off: if it be maimed, or other ways defective, it may be well to cut the plant down to the ground, and train a freſh ſhoot; or, if the head

be

be taken off ſmooth, immediately above a ſtrong ſide ſhoot, this will ſometimes outgrow the crookedneſs, and, in a few years, become a ſtraight plant.

THE time of the plants remaining in the nurſery is determined by a variety of circumſtances; and a ſeaſonable THINNING frequently becomes neceſſary. In this part of the buſineſs there are general rules to go by: the ſhrubby ſpreading tribes ſhould be thinned whenever their branches begin to interfere; and the ſtem plants, whenever their roots get into a ſimilar ſtate. If either the tops of the one, or the roots of the other, be ſuffered to remain in a ſtate of interference and warfare with each other, the beauty of the ſhrubs will ſoon be deſtroyed, and the thriftineſs of the trees will be checked. If the plants be wanted for planting out, it is fortunate; if not, every alternate plant ſhould be moved to a vacant ground, prepared for the purpoſe. If ſuch as ſtand in rows be removed, alternately, into the intervals, and ſet in the quincunx form, a temporary relief will be gained, at a ſmall expence.

PLANTERS, in general, are not aware of the caution neceſſary in TAKING UP plants, for the purpoſe of planting them out to ſtand. *In this caſe*, every root and every fibre ought, as much as poſſible,

poſſible, to be preſerved. No violence ſhould, therefore, be uſed in this operation. The beſt way is to dig a trench cloſe by the ſide of the plant to be taken up; and, having undermined the roots, let the plant fall of itſelf, or with a very little aſſiſtance, into the trench: if any licentious root or roots ſtill have hold, cut them off with ſome ſharp inſtrument, ſo as to jar the main root as little as may be. If the root was properly pruned before planting, it will now turn out a globular bundle of earth and fibres, the beſt characteriſtic of a well rooted plant.

WHEN the nurſery lies at a diſtance from the plantations, or when the plants are to be ſent to ſome diſtant place, much depends upon PACKING them up judiciouſly. Valuable plants may be packed in pots or baſkets;—ſtraw may, however, in general be uſed, and will equally preſerve them from froſt in winter, and the drought of autumn or ſpring; eſpecially if, in the latter caſe, the ſtraw be occaſionally moiſtened with ſoft water. Large plants ſhould be packed, ſingly, with as much earth about the roots as conveniency will allow. If a piece of mat be put over the ſtraw, it will ſave ſome trouble in cording, and be more effectual than ſtraw alone.

SECTION

SECTION THE THIRD.

PLANTING TREES AND SHRUBS.

ALL that we propoſe, in this place, is to convey to our readers ſome general ideas of

PREPARING THE SOIL,	TER PLANTING,
SEASON OF PLANTING,	CLEANING THE PLANTATION,
PRUNING AND SORTING THE PLANTS,	PRUNING THE YOUNG TREES, AND
PLANTING THEM OUT,	THINNING THEM, —
ATTENDING THEM AF-	

in order to avoid uſeleſs repetitions, when we come to ſpeak, ſeparately, of each individual ſpecies; and to enable ſuch of our readers as are wholly unacquainted with the ſubject, to follow us through the ALPHABET OF PLANTS with a degree of eaſe and ſatisfaction to themſelves, which, without theſe previous inſtructions, they would not be able to do.

THE PREPARATION OF THE GROUND depends, in ſome meaſure, upon the ſize of the plants. To ſpeak generally upon the ſubject — For plants under

under four feet high, the ſoil ought to receive a double digging, or a ſummer's fallow under the plow, or a crop of turnips well hoed; but, for larger plants, ſeparate holes, dug in the unbroken ground, are frequently made uſe of; though we cannot, by any means, recommend the practice. Trees and ſhrubs never thrive better, than when they are planted upon *made ground*; for here the fibres rove at large, and the nearer the ſoil of a plantation is brought to the ſtate of made ground; that is to ſay, the more it is broken, and the deeper it is dug; the greater probability there will be of ſucceſs. Plants put in holes may thrive very well, while the fibres have looſe mold to work in; but, whenever they reach the firm unbroken ſides of the pits, they will, except the ſoil be of a very rich loamy nature indeed, receive a check, which they will not overcome, for many years. The ſize of the holes, whether in broken or unbroken ground, muſt be in proportion to the roots of the plants to be put in. For large nurſery plants, the holes, in unbroken ground, ſhould not be leſs than two feet deep; and for plants from four to eight feet high, the holes ought to be made from two to four feet diameter: the different ſtrata ſhould be kept ſeparate; laying the ſod on one ſide of the hole, the corn mold or ſoil on another, and the ſubſtratum on a third; and in this ſtate they ſhould lie ſome weeks, before the time of planting.

THIS,

THIS, namely the TIME OF PLANTING, varies with the ſpecies of plant, and with the nature of the ſoil. Plants, in general, may be ſet out either in the autumn, or in the ſpring. In a bleak ſituation, the latter is generally preferable; provided the planting be not done too late. The latter end of February, and all March, is a very proper ſeaſon for moſt plants: but where the ſcene of planting is extenſive, every fit of open weather, during the ſix winter months, ſhould be embraced. Some plants, however, are partial to particular ſeaſons: theſe peculiarities will be mentioned, in their proper places.

IT has been already intimated, that, when trees and ſhrubs are planted out finally, their *roots* ſhould be left UNPRUNED. It is uſual, and may be proper, to take off the bruiſed and maimed parts; but even this ſhould be done with caution. Their *tops*, however, require a different treatment. Foreſt trees, and other ſtem plants, may in general be trimmed cloſely; by which means the roots will be able to ſend up a ſufficient ſupply of nouriſhment and moiſture the firſt year, and thereby ſecure the life of the plant: whereas, on the contrary, if a number of ſide ſhoots be left on, the quantity of leaves and ſhoots becomes ſo great, that the plant probably is *ſtarved*, for want of that neceſſary ſupply. This renders the ſucceſs of

ſhrubby

ſhrubby plants uncertain; and is an argument againſt their paſſing through the nurſery; and, of courſe, in favor of their being moved (when practicable) from the ſeminary into the place in which they are intended to remain. A well rooted plant, however, if planted in a good mold and a moiſt ſeaſon, will ſupport a conſiderable top; and there is a general rule for the pruning of plants: Leave them tops proportioned to their roots; for no doubt the larger the top, provided the root can ſupport it, the quicker progreſs the plant will make: nevertheleſs, it is well to be on the ſafer ſide; a ſure though ſlow progreſs is preferable to a dead plant, which is always a reflection upon the planter, and an unſightly incumbrance in the plantation. A judicious planter, while he trims his plants, will at the ſame time SORT them: inſtead of throwing them out of his hand into one heap, promiſcuouſly, he will lay the weak ill rooted plants, in one place; the middle ſort, in another; and the ſtrong well rooted ones, in a third; in order that, when they are planted out, each plant may have a fair and equal chance of riſing; which, without this precaution, cannot be the caſe.

WE now come to the operation of PLANTING; which is guided, in ſome meaſure, by the ſpecies of plantation. If the plants be large, and the plantation chiefly ornamental, they ought to be planted out promiscuously in the situation in which they are intended to remain; but, if the plants be small, and the plantation chiefly useful, nursery rows ought generally to be preferred. For, in this manner, the tender plants give warmth to each other; the transition is less violent, than when they are planted out immediately from the nursery or seedbed, singly, and at a distance from each other: the ground is more easily kept clean, than where the plants stand in the random manner; besides, the intervals may, while the plants are young, be cropped with advantage: while the remainder of the intended plantation may be kept in an entire state of cultivation, until the plants acquire a considerable size; or, if the whole ground be stocked in this nursery manner, the superfluous plants may, in almost any country, be sold to great profit. We do not recommend planting these *nursery plantations* too thick; four feet between the rows and two feet between the plants are convenient distances; or, if the intervals be set out exactly a quarter of a rod wide, namely, four feet one inch and a half, and the plants be put in at twenty four inches and three quarters apart, the calculation of how many plants will be required for an acre, or any other given portion of ground, or, on the contrary, how much ground will be necessary for a given number of plants, will be made easy and certain. The method of putting in the

the plants, in theſe nurſery rows, is this: The ground being brought to a proper ſtate of cultivation, as directed above, the plants trimmed and ſorted, and the rows ſet out, a line is laid along, to make the holes by. To aſcertain preciſely the center of each hole, a mark is made in the line (or land-chain, which is not liable to be varied in its length by the weather), and a ſtick, or other guide, placed where the center of each hole falls. The workmen begin to make the holes, by chopping a ring round each ſtake, with the ſpade, of a diameter proportioned to the ſize of the plants, and of a depth equal to that of the cultivated mold. A row of holes being finiſhed, the plants, in this caſe, may be immediately put in; which is done in this manner: One man, or boy, holds the plant upright, with its ſtem in the center of the hole, at the ſame time looking along the row, to ſee that it ſtands in its proper line, while another fills in the mold; firſt ſpreading the roots and fibres level in the bottom of the hole; being careful not to ſuffer any of them to lie in a cramping folded ſtate; but opening them wide, and ſpreading them abroad in the manner of a bird's foot. While the planter is bedding the roots in the fineſt of the mold, the perſon who ſteadies the plant ſhould move it very gently up and down, if ſmall, but if larger, by a

 cir-

circuitous motion of the top, in order to let in the mold more effectually among the fibres; which done, they ſhould be preſſed down gently together with the foot; and the treading, if the ſoil be light, ſhould be repeated two or three times, until the hole be filled up round, and the plant firmly fixed, at the ſame depth at which it ſtood, in the place from whence it was taken. If, on trial, the hole be found too ſhallow, it muſt be deepened; if too deep, ſome of the rougheſt of the mold muſt be thrown to the bottom, until the roots be brought to their natural level. The row being finiſhed, the planter walks back along it, and adjuſts ſuch plants as lean or ſtand out of the line, while his helper diſtributes the plants of the next row. In a ſimilar manner the plants are put in, when the holes are made in whole ground. The ſods are generally thrown to the bottom of the hole; and, if theſe be not ſufficient to raiſe the plant high enough, ſome of the ſubſtratum is mixed with them; or if this be of a very bad quality, ſome of the top ſoil is dug from the intervals, and thrown into the hole. The roots are bedded in the beſt of the mold, and the hole rounded up, either with the ſubſtratum or with the ſoil of the interval, ſo as to form a hillock or ſwell round the ſtem of the plant, in order to allow for the ſettling of the broken mold.

PLANTATIONS require a cloſe ATTENDANCE AFTER PLANTING; eſpecially in windy weather. Large plants are generally ſtaked; but this is a practice we do not recommend, except for large tranſplanted trees or ſhrubs: but of theſe in the next Section. Plants, even of ſix or eight feet high, if well rooted and *firmly* planted, will withſtand a great deal of weather. The plantation, however, ſhould frequently be gone over, and ſuch plants as have loſt their upright poſture, or are looſe at the roots, ſhould be righted and *rammed*; eſpecially if the ſoil be of a light open texture: indeed, in ſuch a ſoil, it is prudent to ram them, at the time of planting; which not only prevents their being miſplaced by the winds, but alſo prevents the drought from reaching the roots ſo ſoon as when the mold is left light and porous. In this caſe, however, it muſt be remembered, that when the plants have got foothold, the mold which has been rammed ſhould be looſened with the ſpade; in order to admit a full ſupply of air to the roots, without which no plant can flouriſh. If a continuance of drought ſet in, after planting, it will be prudent to WATER the plants *; not partially, by pouring a ſmall quantity of water againſt the ſtem of each; but in large quantities, poured

* Provided the HOLES have not been previouſly WATERED; a precaution which, in a dry ſoil and a dry ſeaſon, ought not to be omitted.

into a ring made near the outſide of the hole; ſo that the whole maſs of broken earth may be thoroughly moiſtened, without waſhing off the finer mold from the fibres. A ſuperficial watering tantalizes the plants, and leads the fibres towards the ſurface for nouriſhment: the moiſture, thus partially given, ſoon evaporates, and the diſappointed fibres become expoſed to the parchings of the ſun and wind.

PLANTATIONS in rows are beſt CLEANED with the plow. In the ſpring, gather two furrows, or if room four furrows, into the middle of the interval: in ſummer, ſplit theſe interval ridges, throwing the mold to the roots of the plants, to ſave them from the drought: in autumn, gather them again into the interval; and in winter, again return them to the rows, to keep out the froſt. If the ſoil be good, and dung can be had, a row of potatoes, cabbages, &c. may be planted in each interval, or turnips ſown over the whole: in either caſe, the interſpaces of the rows ſhould be kept clean hoed, or hand-weeded. In like manner, a promiſcuous plantation ſhould be dug, or turned in, at leaſt once a year, for three or four years after planting.

As the plants increaſe in bulk and ſtature, they will require PRUNING. Much depends upon doing

doing this judiciously. If it has been neglected too long, care must be had not to do too much at once. The leader is the principal and first object; the side branches may be afterwards taken off gradually, so as not to wound the plant too much, nor let too much air, at once, into the plantation. The time of pruning is generally considered to be in autumn or winter, when the sap is down, and the leaves off; but, for plants which are not liable to *bleed*, we rather recommend midsummer; as shoots taken off at that time, are not so apt to be followed by fresh shoots, as those taken off in winter. If the shoots be young and slender, it is better to *rub* them off, than to cut them off, clean, with a sharp instrument; boughs and strong shoots, however, require an instrument; and, from young trees, they should be taken off as smooth and close to the stem as possible. If a stump be left, it will be some years before it be grown over, and a flaw, if not a decayed place, will be the consequence; but if a shoot, or even a considerable bough of a young growing tree, be taken off, level with the bark of the stem, the wound will skin over the first year, and in a year or two more no traces of it will be left. A large bough of an aged full-grown tree requires a different treatment; which will be given under the article HEDGEROW TIMBER.

GREAT judgment is required in THINNING plantations. The ſame rule holds good in nurſery plantations, as in the nurſery itſelf, and the ſame general rule (liable no doubt to many exceptions) may be extended to woodlands, and ornamental plantations. But of theſe hereafter: ſuffice it to repeat, in this part of our Work, that whenever the roots of plants begin to interfere with each other, their growth, from that time, is more or leſs checked; and whenever their branches are permitted to claſh, from that time their beauty and elegance are more or leſs injured.

SECTION THE FOURTH.

TRANSPLANTING TREES AND SHRUBS.

BY this is meant the removal of trees and ſhrubs, which, having formerly been planted out, have acquired ſome conſiderable ſize. We do not mean to recommend the practice, in general terms; but, for thinning a plantation, for removing obſtructions, or hiding defects, or for the purpoſe of raiſing ornamental groups or ſingle trees expeditiouſly, it may frequently be uſeful, and is univerſally practiſed; though ſeldom with uniform ſucceſs. This is, indeed, the moſt difficult part of planting,

planting, and requires conſiderable ſkill—with great care and attention in applying it.

IT is in vain to attempt the removal of a tap-rooted plant (as the oak), which has not previouſly been *tapped*; that is, its tap root taken off; and not leſs arduous to make a weakly rooted plant, of almoſt any ſpecies (the aquatics excepted), ſucceed with a large top upon it; much, therefore, depends upon taking up, and pruning, trees and ſhrubs for tranſplantation.

BEFORE a tap-rooted plant, which has never been removed from its place of ſemination, can be taken up with propriety, it muſt be tapped in this manner: Dig a trench or hole by the ſide of the plant, large enough to make room to undermine it, in ſuch a manner as to be able to ſever the tap-root; which done, fill in the mold, and let the plant remain in this ſtate one, two, or three years, according to its ſize and age. By this time the horizontal roots will have furniſhed themſelves with ſtiength and fibres; eſpecially thoſe which were lopped in the excavation; and the plant may be taken up and removed, in the ſame manner as if it had been tapped and tranſplanted while a ſeedling, though not with equal ſafety; for plants that have never been removed, have long branching roots, and the *fibres* lie at a diſtance from ſrom the body of the plant; while thoſe which have been taken up, and have had their roots trimmed when young, are provided with fibres, which, being leſs remote from the ſtem, may be taken up with the plant, and conveyed with it to its new ſituation. This naturally leads to what may, perhaps, be called a *refinement*, in taking up large fibrous-rooted plants for tranſplantation; namely, lopping the whole, or a part of the horizontal roots, two years, or a longer time, before the plant be taken up; leaving the downward roots, and (if neceſſary) part of the horizontal ones, to ſupport the plant until the time of removal *. It would be needleſs to add, that in taking up plants, in general, the greater length of root, and the greater number of fibres there is taken up, the more probable will be the ſucceſs. It is alſo a circumſtance well underſtood, that too much earth cannot be retained among the fibres †.

THE plant being thrown down, and the roots diſentangled, it is proper, before it be removed from its place, to prune the top, in order that the carriage may be lightened. In doing this, a

* In this caſe the head ought, at the ſame time, to be pruned, and the plant, if expoſed, to be ſupported.

† But ſee MINUTES 12 and 15.

con-

considerable ſhare of judgment is requiſite: to head it down in the pollard manner, is very unſightly; and to prune it up to a mere maypole, or ſo as to leave only a ſmall broom-like head at the top, is equally deſtructive of its beauty. The moſt rational, the moſt *natural*, and, at the ſame time, the moſt elegant, manner of doing this, is to *prune the boughs*, in ſuch a manner as to form the head of the plant into a conoid, in reſemblance of the natural head of the Lombardy poplar, and of a ſize proportioned to the *ability of the root*. Whoever was the inventor of this method of pruning the heads of trees, deſerves infinite credit: it only wants to be known in order to be approved; and we are happy to ſee it growing into univerſal practice.

THE mode of carriage reſts wholly with the ſize of the plant: if ſmall, it is beſt carried by hand, either upon the ſhoulder, or upon handſpikes *:—if larger, two ſledges, one for the root, the other for the head, may be uſed:—if very large, and the ball of earth be heavy, a pair of high timber wheels (guarded by a ſack of hay or other ſoft ſubſtance), or a timber carriage, will be found neceſſary.

* See MINUTES, as above.

THE

THE hole muſt be made wide enough to admit the root of the plant, with a ſpace of a foot, at leaſt, all round it, for the purpoſe of filling in the mold with propriety; ſo that if the tree was taken up with a root of eight feet diameter, the hole muſt be made of the diameter of nine or ten feet, and of a depth ſufficient to admit of the tree's being ſeated (when the mold is ſettled) at its natural depth, as alſo to receive the ſods, and other rough unbroken mold, at leaſt a foot thick underneath its root.

THE method of planting depends upon the ſtate of the root, and the temperature of the mold and the ſeaſon. If the root be well furniſhed with fibres and mold, and the ſoil be moiſt from ſituation, or moiſtened by the wetneſs of the ſeaſon, no artificial preparation is neceſſary. The bottom of the hole being raiſed to a proper height, and the tree ſet upright in the center of it, the mold may be filled in; being careful to work it well in amongſt the roots, and to bed the fibres ſmoothly amongſt it; treading every layer firmly, and, with a carpenter's rammer, filling every crevice and vacancy among the roots, ſo that no ſoft part nor hollowneſs remain; and proceed in this manner, until the hole be filled, and a hillock raiſed round the plant to allow for its ſettling. But if the roots be naked of mold, and thin of fibres, and

and the ſoil, the ſituation, and the ſeaſon be dry, we recommend the following method: The requiſite depth of the hole being aſcertained, and its bottom raiſed to a proper height, with ſome of the fineſt of the mold, pour upon it ſo much water as to moiſten the looſe mold, without rendering it ſoft, and unable to ſuſtain the weight of the plant; and th n proceed as above directed. If the tranſplantation be done in autumn, it will require nothing farther at that time; but if in the ſpring, more water will immediately be wanted. Therefore, at once, draw a ring, ſome inches deep, near the outſide of the hole, and, in the bottom of its channel, make ſix, eight, or ten holes (by means of an iron crow, or of a ſpike and beetle), at equal diſtances, and of a depth equal to that of the roots of the plant. Theſe holes will not only ſerve to convey water, but air alſo, to the immediate region in which they are both indiſpenſably neceſſary to the health of the plant. We have been the fuller in our inſtructions relative to tranſplanting, as being a proceſs little underſtood by profeſſional men. Every nurſeryman, and almoſt every kitchen gardener, can raiſe, train, and plant out ſeedling and nurſery plants; but the removal of trees ſeldom occurs in their practice; and we have met with very few men, indeed, who are equal to the taſk. The foregoing rules are the reſult of experience.

FOR

FOR farther experience in TRANSPLANTING, see MINUTES 12 and 15, in this Volume. And for farther remarks on Planting in general, see The RURAL ECONOMY of the MIDLAND COUNTIES, VOL. II. MINUTES 146 and 168.

DIVISION

DIVISION THE SECOND.

CHOICE OF TIMBER TREES.

SECTION THE FIRST.

CONSUMPTION OF TIMBER.

TIMBER is the great and primary object of planting. Ornament, abstracted from utility, ought to be confined within narrow limits. Indeed, in matters of planting, especially in the taller plantations, it were difficult to separate, entirely, the idea of ornament from that of use. Trees, in general, are capable of producing an ornamental effect; and there is no tree which may not be said to be more or less useful. But their difference in point of value, when arrived at maturity, is incomparable; and it would be the height of folly to plant a tree whose characteristic is principally ornamental, when another, which is more useful and equally ornamental, may be planted in its st

THERE-

THEREFORE, previous to our entering, at large, upon the businefs of planting, it will be proper to endeavour to fpecify the trees moft ufeful to be planted. In attempting this, we muft look forward, and endeavour to afcertain the fpecies and proportional quantities of TIMBER which will hereafter be wanted, when the trees, now to be planted, fhall have reached maturity. To do this with a degree of certainty, is impoffible: Cuftoms and fafhions alter, as caprice and neceffity dictate. All that appears capable of being done, in a matter of this nature, is, to trace the great outlines, and, by obferving what has been permanently ufeful for ages paft, judge what may, in all human probability, be ufeful in ages to come.

SHIPS, | MACHINES, and
BUILDINGS, | UTENSILS,

have been, are, and moft probably will continue to be, the confumers of TIMBER, in this country. We will, therefore, endeavour to come at the principal materials made ufe of in the conftruction of thefe four great conveniences of life. Indeed, while mankind remain in their prefent ftate of civilization and refinement, they are *neceffaries* of life, which cannot be difpenfed with; and are confequently objects which the planter ought not to lofe fight of, as they include, in effect, every thing that renders plantations ufeful; FENCE WOOD and FUEL excepted.

I. SHIPS

I. SHIPS are built chiefly of OAK: the keels, however, are now pretty generally laid with ELM, or BEECH; and part of the upper decks of men of war is of DEAL: but theſe woods bear no proportion, in reſpect of the quantity uſed, to the Oak. The *timbers* of a ſhip are principally crooked, but the *planking* is cut out of ſtraight pieces. In a ſeventy-four gun ſhip, the crooked and ſtraight pieces uſed are nearly equal, but the *planking under water* is of FOREIGN OAK: therefore, of ENGLISH OAK, the proportion of crooked to ſtraight pieces is almoſt two to one. Maſts and yards are of DEAL. The blockmakers uſe Elm, Lignum-Vitæ, Box, and other hard woods. Upon the whole, it may be ſaid, that, in the conſtruction of a ſhip, OAK is the only ENGLISH WOOD made uſe of; and that, of this Engliſh Oak, nearly two thirds are requiſite to be more or leſs CROOKED.

II. BUILDINGS. In the metropolis, and towns in general, DEAL is the prevailing wood made uſe of by the *houſe carpenter*: ſome OAK is uſed for ſaſhes, alſo for window and door frames, and ſome for wall plates; but in places ſituated within the reach of water carriage, DEAL is becoming every day more and more prevalent: nevertheleſs, there are many inland parts of the country, where the houſe carpenters ſtill continue to work up great quantities of OAK and ELM. The *joiner*

 ſcarcely

ſcarcely uſes any other wood than DEAL, except in ſome inland and well wooded diſtricts, where OAK is ſtill in uſe for floors and ſtaircaſes. Through the kingdom at large, perhaps three fourths of the timber uſed in the conſtruction of buildings are FOREIGN DEAL

III. MACHINES. This claſs comprehends MILLS and other MACHINES of MANUFACTORY, CARRIAGES of burden and pleaſure, IMPLEMENTS OF HUSBANDRY, with the other articles neceſſary in rural affairs.

THE *millwright*'s chief material is OAK, and ſome CRABTREE, for cogs *.

THE *waggon and cartwright* uſes OAK, for bodies; ASH, for ſhafts and axles; ELM, for naves, and ſometimes for fellies and linings.

THE *plowright*'s ſheet anchor is ASH: in ſome counties BEECH is ſubſtituted in its ſtead, for every thing but plow beams.

THE *coachmakers* uſe ASH, for poles, blocks, ſplinter bars, &c. ELM, for naves; generally

* As to the implements, utenſils, and machines of manufactory, they are infinite; and various kinds of wood are worked up in making them.

ASH,

ASH, for spokes and fellies; and RATTAN *, for pannels.

Gates and Fences are made of OAK and DEAL; sometimes of ASH, ELM, MAPLE, &c. but *posts* are, or ought to be, universally of OAK, CHESNUT, or LARCH;

Ladders, of DEAL, OAK, &c.

Pumps and Water Pipes, of OAK, ELM, ALDER;

Wooden Bridges, River Breaks, and other *Waterworks*, principally OAK; some ELM and ALDER under water †.

IV. UTENSILS. Under this head we class FURNITURE, COOPER'S WARE, MATHEMATICAL INSTRUMENTS, TRUNKS, PACKING CASES, COFFINS, &c. &c.

THE *cabinetmakers'* chief woods are MAHOGANY and BEECH; next to these follow DUTCH OAK (Wainscot), DEAL, ELM; and lastly,

* The mahogany of the Bahama Islands.

† BEECH has lately been found to lie long under water; but for *waterwork* of every kind the LARCH is found to excel.

WALNUTTREE, CHERRYTREE, PLUMTREE, BOX, HOLLY, YEW, and a variety of woods for inlaying and cabinets. In some country places, a considerable quantity of ENGLISH OAK is worked up into tables, chairs, drawers, and bedsteads; but, in London, BEECH is almost the only English wood made use of, *at present*, by the cabinet and chair makers.

THE *carvers'* favorite wood is LIME, for picture and glass frames; DEAL, for coarser articles.

Coopers;—OAK (and some CHESNUT), for large casks and vessels: ASH, for dairy utensils, butter firkins, flour barrels, &c. OAK, for well buckets and water pails, and, in some places, for milk pails, and other dairy vessels: BEECH, for soap firkins, &c.

Locksmiths, in Birmingham and Wolverhampton, work up a considerable quantity of OAK, for Lockstocks: chiefly the butts of trees.

Turners;—principally BEECH for large ware, if BEECH is to be had; if not, SYCAMORE, or other clean-grained wood: BOX, HOLLY, &c. for smaller utensils.

Mathematical Instrument Makers;—MAHOGANY, BOX, HOLLY.

Trunk-

Trunkmakers ;—DEAL.

Packing Cases ;—also DEAL.

Coffins ;—OAK, ELM, DEAL.

AND, finally, the *laſtmakers*, who work up no inconſiderable quantity of wood, uſe BEECH for laſts; ALDER and BIRCH for heels, patten-woods, &c.

WE do not deliver the foregoing ſketch as a perfectly correct account of the application of woods, in this country: the attempt is new, and that which is new is difficult. We have not omitted to conſult with profeſſional men upon the ſubject; and we believe it to be ſufficiently accurate for the purpoſe of the planter. If we have committed any material error, we aſk to be ſet right. We do not wiſh to deſcend to minutiæ: it would be of little uſe to the planter, to be told what toys and toothpicks are made from: it is of much more importance to him to know, that, of ENGLISH WOODS, the OAK is moſt in demand, perhaps three to one,—perhaps in a much greater proportion; that the ASH, the ELM, the BEECH, and the BOX, follow next; and that the CHESNUT, the WALNUT, and the PRUNUS and PINUS tribes, are principally valuable, as ſubſtitutes for OAK and FOREIGN TIMBER.

SECTION THE SECOND

SITUATION AND SOIL.

IN the choice of timber trees, however, SITUATIONS and SOILS muſt ever be conſulted. The *Oak*, in ſhallow barren ſoils, and in bleak expoſed ſituations, cannot be raiſed with profit, as a timber tree; while the *Larch*, by out-growing its ſtrength, ſickens in deep rich ſoils.

IT is a fortunate circumſtance for this country, that the two trees which are moſt likely to furniſh its navy with an internal ſupply of timber, ſhould delight in ſoils and ſituations of oppoſite natures; and every judicious planter will endeavour to aſſign to each its natural ſtation.

DIVISION

DIVISION THE THIRD.

HEDGES AND HEDGEROW TIMBER

INTRODUCTORY REMARKS.

THE raiſing of LIVE HEDGES and HEDGEROW TIMBER conſtitutes no inconſiderable part of the buſineſs of planting. The value of good Hedges is known to every huſbandman; and notwithſtanding the complaints againſt Hedgerow Timber, as being liable to be knotty, &c. the *quality* of the timber itſelf is not queſtioned: its faultineſs ariſes, wholly, from an improper treatment of the tree, and not from the ſituation of its growth. Indeed we are clear in our opinion, that, under proper management, no ſituation whatever is better adapted to the valuable purpoſe of raiſing SHIP TIMBER, than Hedges: The roots have free range in the adjoining incloſures, and the top is expoſed to the exerciſe of the winds, with a ſufficient ſpace to throw out luſty arms, and form, at a proper height, a ſpreading head. Thus, quickneſs of growth, with ſtrength and CROOKEDNESS of Timber, are at once obtained.

WE are well aware of the injury reſulting from woody Hedgerows to arable incloſures; but every man experienced in rural matters muſt be convinced, that it is not well trained Timber trees, but high Hedges, and low Pollards, which are the bane of corn fields. Theſe, forming a high and impervious barrier, preclude the air and exerciſe, ſo eſſential to the vegetable, as well as the animal creation: in Norfolk, lands thus encumbered are, with great ſtrength and propriety of expreſſion, ſaid to be *wood-bound*. Beſides, Pollards and low-ſpreading trees are certain deſtruction to the Hedge wood which grows under them.

NEITHER of theſe evils, however, reſult from tall Timber Oaks, and a Hedge kept down to four or five feet high: a circulation of air is, in this caſe, rather promoted than retarded; and it is well known, that a pruned Hedge will thrive perfectly well under tall-ſtemmed trees, Oaks more eſpecially. We will therefore venture to recommend, for arable incloſures, Hedges pruned down to four or five feet high, with Oak timbers of fifteen to twentyfive feet ſtem.

BUT

BUT, for graſs lands, higher Hedges are more eligible. The graſſes affect warmth, which promotes their growth, and thereby increaſes their quantity, though their quality may be injured. Beſides, a tall fence affords ſhelter to cattle; provided it be thick and cloſe at the bottom; otherwiſe, by admitting the air in currents, the blaſt is rendered ſtill more piercing. The ſhade of trees is equally friendly to cattle in ſummer, as thick Hedges are in the colder months; therefore, in the Hedges of graſs incloſures, we wiſh to ſee the Oak wave its lofty ſpreading head, while the Hedge itſelf is permitted to make its natural ſhoots: remembering, however, that the oftener it is cut down the more durable it will be as a fence, and the better ſhelter it will give to cattle; more eſpecially if the ſides be pruned the firſt and ſecond years after cutting, in order to give it an upright tendency, and thicken it at the bottom.

UPON bleak hills, and in expoſed ſituations, it is well to have two or even three rows of Hedge wood, about four feet apart from each other; the middle row being permitted to reach, and always remain at, its natural height; while the ſide rows are cut down, alternately, to give perpetual ſecurity to the bottom, and afford a conſtant ſupply of materials for Dead hedges, and other purpoſes of Underwood.

HAVING

HAVING thus given a general ſketch of our ideas as to the different kinds of Hedges, and their effects on cultivated lands. we proceed to treat of the method of raiſing them. In doing this, it will be proper to conſider,

1. THE Woods moſt eligible for Hedges.

2. THE time and manner of planting them.

3. THE manner of defending the young plants.

4. THE method of cleaning and training them.

5. THE after management.

SECTION THE FIRST.

SPECIES OF HEDGE WOODS.

THE SPECIES OF HEDGE WOOD depends, in ſome meaſure, upon ſoil and ſituation. That which is proper for a ſound ſoil, in a temperate ſituation, may not be eligible for a marſh, or a mountain: and, indeed, a fence may be formed of any tree or ſtrong ſhrub, provided it will thrive in the given ſituation. Nevertheleſs, there are ſome ſpecies much

much more eligible than others; we particularize the following:

THE HAWTHORN.
THE CRAB BUSH.
THE AQUATIC TRIBE.
THE HOLLY.
THE FURZE.

THE HAWTHORN has been confidered, during time immemorial, as the wood moft proper for live fences. This pre-eminence, probably, arofe from the feedling plants being readily collected, in woods and waftes; the method of raifing them, in feed beds, being formerly, and indeed in fome parts of the kingdom even to this day, but little practifed. The longevity of the Hawthorn, efpecially if it be frequently cropped, and its patience in cropping,—its natural good qualities as a live fence, and its ufefulnefs as affording materials for dead hedges, are other reafons why it has been univerfally adopted. Another advantage of the Hawthorn—It will grow in almoft any foil, provided the fituation be tolerably dry and warm. However, if the foil hath not a degree of richnefs in itfelf, as well as a geniality of fituation, the Hawthorn will not thrive fufficiently, nor make a progrefs rapid enough, to recommend it, in preference, as a Hedge wood.

THE

THE CRABTHORN, among the deciduous tribe, ſtands next: indeed, taken all in all, it may be ſaid to rival the Hawthorn itſelf. Its growth is conſiderably quicker, and it will thrive in poorer ſoils, and in bleaker ſituations; and although it may not be ſo thorny and full of branches as the Haw-buſh, yet it grows ſufficiently rugged to make an admirable fence. Add to this, though its branches may not be preferable to thoſe of the Hawthorn for *ſhooting* dead hedges, they undoubtedly afford a much greater quantity of *ſtakes*; and no wood whatever (the Yew perhaps only excepted) affords better ſtakes than the Crab tree. The ſeedling plants, too, are readily raiſed, as the ſeeds of the Crab vegetate the firſt year. We do not mean, however, to force down the Crab buſh upon our Reader as being, in a general light, preferable to the Hawthorn: we wiſh only to ſtate, impartially, their comparative value; leaving him to conſult his own ſituation and conveniency, and, having ſo done, to judge for himſelf. Nevertheleſs, from what has been adduced, we may venture to conclude, that upon a barren ſoil, and in a bleak ſituation, the Crab buſh, as a Hedge wood, claims a preference to the Hawthorn.

THE AQUATICS. As the Crab excels the Hawthorn, upon bleak barren hills, ſo the Aquatics gain a preference, in low ſwampy grounds: for although the

the Hawthorn delights in a moiſt ſituation, yet much ſtagnant water about its roots is offenſive to it. Of the Aquatics, the *Alder* ſeems to claim a preference; its growth is more forked and ſhrubby than that of the Poplar or Willow; and its leaves are particularly unſavory to cattle. In point of ornament, however, it is exceeded by the *Black Poplar*, which, if kept pruned on the ſides, will feather to the ground, and form a cloſe and tolerably good ſence.

THE HOLLY. Much has been ſaid, and much has been written, of the excellency of Holly hedges: nevertheleſs, as fences to farm incloſures, they ſtill exiſt in books and theory only; not having yet been introduced into general practice, we believe, in any part of the kingdom. Their ſuperiority, however, whether in point of utility or ornament, is univerſally acknowledged. The perpetual verdure they exhibit, the ſuperior kind of ſhelter they afford, during the winter months, and the everlaſtingneſs of their duration (an old decayed Holly being an object rarely to be ſeen in nature), all unite in eſtabliſhing their excellency. How then are we to account for the ſcarcity of Holly hedges? The difficulty of raiſing them, and the ſlowneſs of their growth, have been held out as obſtacles; and ſuch they are, in truth; but they are obſtacles ariſing rather from a want of proper

proper management, than from any caufe inherent in the Holly itfelf. Thoufands of young Hollies have been deftroyed, by being planted out improperly, in the fpring, at the time that the Hawthorn is ufually planted: and the few which efcape total deftruction, by fuch injudicious removal, receive a check which cripples their growth, probably for feveral years.

We do not mean to intimate, that, by any treatment whatever, the progrefs of the Holly can be made to keep pace with that of the Hawthorn, or the Crab: and we are of opinion that it ought, by reafon of the comparative flownefs of its growth, to be raifed *under* one or other of thefe two plants; more efpecially under the Crab, which, as has been obferved, has a more upright tendency than the Hawthorn, and confequently will afford more air, as well as more room to the Holly rifing under it.

But whilft we thus venture to recommend raifing the Holly under the Crabthorn, we are by no means of opinion that it is difficult to raife a hedge of Holly alone. The principal difadvantage arifing comparatively from this practice is, that the dead fence will be required to be kept up at leaft ten or twelve years, inftead of fix or feven; in which time a Crabthorn hedge, properly managed, may be made a fence, and will remain fo, without

without further expence, until the Holly become impregnable; when the Crab may either be removed, or permitted to remain, as taste, profit, or conveniency may point out.

THE Holly will thrive upon almost any soil; but thin-soiled heights seem to be its natural situation. We may venture to say, that where corn will grow, Holly will thrive abundantly; and Holly hedges seem to be peculiarly well adapted to an arable country: for, being of slow growth, and its perspiration being comparatively small, the Holly does not *suck the land* (as the Countryman's phrase is), and thereby rob the adjoining corn of its nourishment, so much as the Hawthorn; which, if suffered to run up to that unpardonable height, and to straggle abroad to that shameful width, at which we frequently see it, is not much less pernicious, in its effects upon corn land, than the Ash itself.

THE FURZE is rather an assistant Hedge wood, than a shrub which, alone, will make a fence. Upon light barren land, however, where no other wood will grow to advantage, tolerable fences may be made with Furze alone.

THERE is one material disadvantage of Furze, as a live Hedge wood; the branches are liable to be killed

killed by ſevere froſts, eſpecially if the plants be ſuffered to grow tall, branchy, and thin at the bottom. It follows, that the beſt preſervative againſt this malady is, to keep them cropped down low, and buſhy; indeed, they are of little uſe, as a fence, unleſs they be kept in that ſtate.

In Norfolk, it is a practice, which of late years has become almoſt univerſal, to ſow Furze ſeed upon the top of the ditch bank; eſpecially when a new Hedge is planted. In a few years, the Furzes get up, and become a ſhelter and defence to the young quick; and, aſſiſted by the high ditch bank prevalent in that country, afford a comfortable ſhelter to cattle in winter; beſides ſupplying, at every fall, a conſiderable quantity of Farmhouſe fuel.

SECTION THE SECOND.

METHOD AND TIME OF PLANTING HEDGEROWS.

I. THE FENCEWOOD. The method varies with the ſoil, and the time with the ſpecies of wood to be planted.

In

IN a low level country, ditches become uſeful, as main drains to the adjoining incloſures; but, in a dry upland ſituation, drains are leſs wanted; and here the Planter has it in choice, whether he will plant with or without a ditch.

THE prevailing cuſtom, taking the kingdom throughout, is to plant with a ſhallow ditch, laying the plants in a leaning poſture againſt the firſt ſpit turned upſide down, covering their roots with the beſt of the cultivated mold, and raiſing a bank over them, with the remainder of the excavated earth of the ditch, without any regard being had to the wetneſs or dryneſs of the ſituation. It is a ſtriking fact; indeed, that in the vale of Glouceſter— where large plots of naturally rich land are chilled with ſurface water, and reduced to little value, entirely thro' a want of proper ſewers and ditches— it is the cuſtom to plant Hedges with a paltry grip of twelve to fifteen inches deep; while in Norfolk —a dry ſandy country, where the natural abſorbency of the ſubſtratum is ſeldom or ever ſatiated —it is the univerſal practice to raiſe Hedges with what is there called a "ſix-foot dyke;" and, when freſh made, they frequently run from ſix and a half to ſeven feet; meaſuring from the bottom of the ditch to the top of the bank.

WHAT may appear equally extraordinary, to the reſt of the kingdom, the Norfolk Huſbandmen, inſtead of planting the quick at the foot of the bank, among the corn mold, lay it in, near the top of their wall-like bank, among the crude earth, taken out of the lower part of the ditch. It is no uncommon ſight, however, in that country, to ſee the face of the bank, with the quickſets it contains, waſhed down, by beating rains, for rods together. Nevertheleſs, if the plants eſcape this accident, it is aſtoniſhing to ſee the progreſs they will ſometimes make, for a few firſt years after planting. But, as the roots enlarge, they become confined for want of room to range in; and the bank naturally moldering down by time, they are left naked and expoſed. It is common to ſee young plants hanging, with their heads downward, againſt the face of the bank; and the mold continuing to crumble away from their roots, they of courſe drop finally into the ditch.

IF we examine the unbroken flouriſhing Hedges of that country, of fifteen, twenty, and thirty years ſtanding (for many ſuch there are, eſpecially in the Fleg Hundreds); we ſhall find them firmly rooted among the corn mold at the foot of the bank. Nevertheleſs, the Norfolk farmers, in general, are ſo cloſely wedded to the foregoing practice,

practice, that no arguments are sufficient to convince them of its impropriety.

WE confess ourselves partial to the superior abilities of the Norfolk Husbandmen, in their general management of rural affairs; and we hold established practices in Husbandry as things too respectable to be wholly condemned without a full and candid examination: we will therefore endeavour, in as few words as possible, to place the Norfolk practice of planting Hedges in its proper light.

THERE are not, generally speaking, any woodlands in Norfolk. The Hedges, it is true, especially of the eastern part of the county, are full, much too full, of wood, chiefly pollards. There are some few timber groves, scattered here and there: but we find none of those extensive tracts of coppice or underwood, in that county, which we see in other parts of the kingdom: consequently, the planter of Hedges experiences a scarcity of materials for temporary dead fences, having neither *stakes*, *edders*, nor *rails*, to make them with. Fortunately for him, however, the *soil* is of such a nature (a light sandy loam of great depth, without a single stone to check the spade), that by digging a deep trench, and raising a mound with the soil, none of those materials are wanted. The face of the bank being carried upright, and a little brushwood set along the top of it, a sufficient fence is formed; while the depth of the ditch prevents cattle from browzing upon the young plants. By this means, Hedges are raised in Norfolk at a trifling expence, compared with the great cost bestowed upon them in some counties; where two rows of posts and rails are used, by way of temporary fences. But the difficulty in raising a live hedge, in the Norfolk manner, arises from the want of a proper place to plant the quicksets in. If it be put in, towards the top of the bank, as is usually done, the evil consequences abovementioned follow: if, on the contrary, it be laid in, near the bottom, the superincumbent pressure of the bank, and the want of moisture in this part of it, render the progress of the young plants slow, for the first three or four years; while those above, having loose *made ground* for their fibres to strike among, and having a sufficient supply of moisture collected from every shower, by the brush hedge, flourish apace; until the roots having grown too large for the bank, or the upper part of the bank itself having been washed down or moldered away, their career is stopt, at a time when those below, having struggled thro' the bank, and finding an ample supply of air, moisture, and rich cultivated soil, to work among, are, in their turn, beginning to thrive; and their main roots being firmly fixed in the soil itself,

itſelf, there is no fear of their afterwards receiving a check.

THUS it appears, that the Norfolk method has its advantage, as being cheap, with a diſadvantage, ariſing from the want of a proper place to put the plants in.

THIS is eaſily obviated by planting with an OFFSET; that is—inſtead of continuing the face of the bank with one unbroken ſlope—to ſet it back a few inches, ſo as to form a break or ſhelf, where the quickſets are planted; for the obvious purpoſe of giving the young plants a ſufficient ſupply of moiſture, air, and paſturage, until their roots have had time to extend themſelves to the adjoining incloſures.

THIS method of raiſing a Hedge is not a mere theoretical deduction, but has been practiſed with ſucceſs, in different parts of the kingdom; and, in a ſoil free from ſtones and other obſtructions of the ſpade, it is perhaps, upon the whole, the moſt eligible practice.

BUT the beſt live hedges we remember to have ſeen, in any part of this kingdom, grow in the neighbourhood of Pickering, in the North Riding of Yorkſhire. Theſe Hedges ſtand nearly upon level

 ground,

ground, with little or no bank or ditch ; ſo that the plants have free range for paſturage, on both ſides ; the ſhallow trenches, by which the quickſets have been planted, being now grown up ; having, it is probable, never been ſcoured out ſince they were made. Indeed, the aſſiſtance of a ditch is not wanted ; no temporary fence whatever being requiſite to be made, when the hedge is topped : the ſtems themſelves are a ſufficient barrier, ſtanding in rows, like the heads of piles, and in ſuch cloſe order, that not a ſheep, nor a hog, nor, in ſome places, even a hare, can creep between them. In a few years, thoſe living piles throw out heads aſtoniſhingly luxuriant, and every ſix or ſeven years afford an ample and profitable crop of bruſhwood ; and this without any expence whatever, except that of reaping it : whereas, in Norfolk, the renewal of the ditch and bank, when the hedge is cut down, is nearly equal to the firſt coſt ; beſides the diſadvantage reſulting from cutting off all communication with the incloſure on the ditch ſide, and thereby robbing the hedge of half its natural food.

THEREFORE, where a ditch is not neceſſary as a drain, and where the nature of the ſubſtratum is ſuch that it cannot be conveniently ſunk ſufficiently deep to defend the young plants—the moſt eligible method, in ſuch a ſituation, is to plant the Hedge upon

upon the LEVEL GROUND, without either bank or ditch, in the manner hereafter to be defcribed; which method is now practifed, in the neighbourhood abovementioned, with very promifing fuccefs.

HAVING thus endeavoured to deduce from actual practice what may be called the *theory* of raifing Hedges, we proceed to the application.

FROM what has been faid, it appears that there are three diftinct methods of raifing a Live Hedge:

1. WITH a ditch and plain bank.
2. WITH a ditch and offset.
3. UPON level ground.

THE *firft* has been already mentioned; and being familiar to every countryman, it is needlefs to enlarge upon it here.

THE *fecond* is to be practifed, in wet fituations, where furface drains are wanted, and where the ditch is neceffary to be kept open; and likewife, in dry fituations, where the fubfoil is fuch that a ditch can be conveniently funk deep enough, to guard the young plants, in front, without an additional fence.

THE manner of executing it is this: The ground may either be prepared by fallowing with the plow, or the work may be lined out upon the unbroken ground. In either caſe, the plants ſhould be ſet *upon* the natural level of the ſoil, and at the diſtance of three to twelve inches from the brink of the ditch. This, in ordinary ſituations, ſhould be about four feet, ſay a quarter of a rod, wide at the top, and being brought to an angle at the bottom (or as near an angle as tools can bring it), its ſlope or ſides ſhould be about the ſame dimenſions; the cavity of the ditch being made, as nearly as may be, an equilateral triangle. But, if the ditch be wanted as a main drain or common ſewer, its width ſhould be conſiderably greater; for, in this caſe, it cannot be *pointed* at the bottom, and muſt therefore have a ſufficient width given it, at the top, to admit of its being made deep enough as a fence, and, at the ſame time, wide enough, at the bottom, to admit the given current of water. The bank ſhould riſe in front, with a ſlope ſimilar to that of the ditch; but as the back ſhould be carried more upright at the foot, ſwelling out full towards the top, in order to admit the inſertion of a bruſh hedge; or, rather, if it can be conveniently had, a dwarf ſtake-and-edder Hedge,

Hedge, which will effectually compleat the fence to the bank ſide; in either caſe, if any ſtraggling ſpray overhang the young plants, it ſhould be trimmed off, with ſome ſharp inſtrument, or be beaten flat with the back of a ſhovel, to prevent its injuring the tender ſhoots.

THE *third* method, namely, planting without a ditch, is more particularly recommended for upland ſhallow ſtony ſoils. In executing this, the ground muſt be previouſly marked out, from four to ſix feet wide, be reduced to a fine tilth, and made perfectly clean, either by a whole ſummer's fallow, repeatedly ſtirred with the plow, or by cultivating upon it, in a huſbandlike manner, a crop of Turnips, or, which is perhaps better, a crop of Potatoes; eſpecially if a little dung can be conveniently allowed them. At the approach of winter, the ſoil being fine and clean, and the crop, if any, off, gather it up into a highiſh round ridge or land, and thus let it lie till the time of planting; when, opening a trench upon the ridge or middle of the land, either with the ſpade or the plow, inſert the plants, upright, filling in the mold, and preſſing it gently to the roots, in the common nurſery manner.

THE ſame precautions ſhould be obſerved, in planting quickſets, that have been already recommended

mended, under the article TRAINING; namely, the plants ſhould be ſorted, as to their ſize, and ſhould be either cut off within a few inches of the ground, or be pruned up to ſingle ſtems.

THE *diſtance* ſhould be regulated by the age and ſtrength of the plants; from four to ſix inches is the uſual diſtance; but if the plants have been previouſly tranſplanted from the ſeed bed, as they ought in general to be, and have acquired four or five years of age and ſtrength, as we would always wiſh they ſhould, from ſix to nine inches is near enough.

THE uſual TIME OF PLANTING is during the ſpring months of February, March, and April; and, for the Hawthorn, the Crab, and the Aquatics, this is at leaſt the moſt convenient ſeaſon; but, for the HOLLY, as will be found under that article, in the ALPHABET OF PLANTS, ſummer is the propereſt time of planting.

WHERE much *ditching* is required, and hands ſcarce, the foundations of the banks may be laid, any time in winter, and left to ſettle, until the time of planting.

THUS far, we have been ſpeaking of raiſing SINGLE HEDGES, whether of Hawthorn, Crab Buſh, or or Holly; we will now ſay a word or two, as to the method of raiſing the HOLLY UNDER THE CRAB or HAWTHORN. This may be done two ways; either by ſowing the berries, when the quick is planted; or by inſerting the plants themſelves, the enſuing midſummer. The firſt is by much the ſimpleſt, and perhaps, upon the whole, the beſt method. The ſeeds may either be ſcattered among the roots of the deciduous plants, or be ſown in a drill in front of them: and if plants of Holly be put in, they may either be planted between thoſe of the Crab, &c. or otherwiſe in front, in the quincunx manner; the tablet of the offset, when a ditch is made uſe of, being left broad for that purpoſe.

IF the FURZE be made uſe of, as an aſſiſtant Hedge wood, it is better to ſow the ſeed on the *back* of the bank, than upon the *top* of it; for, in this caſe, it is more apt to overhang the young plants, in the face of the bank; while, in the other, it is better ſituated, to anſwer the purpoſe intended; namely, that of guarding the back of the bank, as well as of preventing its being torn down by cattle. The method of ſowing the ſeed is this: Chop a drill, with a ſharp ſpade, about two thirds of the way up the back of the bank, making the cleft gape as wide as may be, ſo as not to break off the lip; and having the ſeed in a quart bottle, ſtopt with a cork and gooſe quill, or with a perforated wooden ſtop-

per,

per, trickle it along the drill; covering it by means of a broom, drawn gently above, and over, the mouth of the drill. This is better than closing the drill entirely with the back of the spade, the seeds being sufficiently covered, without being shut up too close, while the mouth of the drill is left open, to receive the rain water which falls on the top of the bank. One pound of seed will sow about forty statute rods. What in Norfolk is called the *French* seed is the best, as the plants from this seldom mature their seeds, in this country; and consequently are less liable to spread over the adjoining inclosure. It may be had at the seed shops, in London, for about fifteen pence a pound.

If a fence be required of Furze alone, a similar drill should be sown on the other side of the bank; and when the plants are grown up, the sides should be cut alternately.

II. Thus much as to planting the Fence; we now proceed to the method of planting HEDGE-ROW TIMBER. It has already been given in opinion, that no situation whatever is better adapted to the raising of ship timber, than Hedges; and we are clearly of opinion, that, in these alone, a sufficient supply, of crooked timbers at least, might be raised, to furnish perpetually the Navy of Great Britain.

Britain. It is a ſtriking fact, that in Norfolk, where there is very little Oak, except what grows in the Hedges, and even in theſe, for one timber tree there are ten pollards, yet the country experiences no want of Oak timber.

But while we recommend the Oak, as eligible to be planted in Hedges, we condemn, as unfit for that purpoſe, every other tree (except, perhaps, the Aquatics in a marſh, the Beech and Pine tribe upon a barren mountain, or the Elm where Oak has lately occupied the ſoil) and more eſpecially the Aſh; not only as being the greateſt enemy to the farmer, but becauſe the excellency of Aſh timber ariſes from a length of ſtem, and cleanneſs of grain: groves, therefore, and not Hedges, are the natural ſituation of the Aſh.

The method of raiſing the Oak in Hedges, may either be by ſowing the acorns, or planting the ſeedlings, at the time of planting the fence wood: we would wiſh to recommend the practice of both; namely, to plant a well rooted thriving nurſery plant (ſuch as has previouſly been tapped and tranſplanted) at the diſtance of every ſtatute rod; and, at the ſame time, to dibble round each plant three or four acorns, to guard againſt a miſcarriage, and to give the judicious woodman a choice in the propereſt plant to be trained.

This

THIS diſtance may be objected to, as being too cloſe; and ſo it may in a deciduous Hedge; but, in a Holly Hedge, we would not wiſh to ſee Oaks ſtand at a greater diſtance; for, ſituated in a Hedge, they have unlimited room to ſpread on either ſide; and, by ſtanding near each other, they are more likely to throw out main branches, fit for ſhip timber, than they would if they had full head room. For this reaſon, it might not be amiſs to plant at every half rod, and, when the Hedge is perceived to begin to ſuffer, to thin them in the manner moſt conducive to the ends propoſed, holding jointly in view the Fence and the Timber.

SECTION THE THIRD.

DEFENDING THE YOUNG PLANTS.

LITTLE more remains to be ſaid upon this head. The ditch, bank, and dwarf hedge have already been fully deſcribed; and this is by much the cheapeſt, and a very effectual, method, where it can be conveniently practiſed; but where the nature of the ſoil is ſuch, that a ditch ſufficiently deep, to defend the young plants, cannot be ſunk but at too great an expence, ſome other expedient muſt be ſought for.

POSTS

POSTS and rails, wound with buſhes in the York-ſhire manner, are an effectual fence; but they are expenſive in the extreme.

IN Surrey and Kent, the prevailing practice is to ſet a ſtrong ſtake-and-edder Hedge behind the quickſets, and throw rough buſhes into a ſhallow ditch, in front: this in a coppice-wood country may be done at a reaſonable expence; but it is by no means effectual.

IN ſome places, wattle Hedges are uſed; and in others furze faggots, ſet in cloſe order, are found effectual, for this purpoſe: in ſhort, almoſt every country affords its own peculiar materials, and every judicious planter will endeavour to find out thoſe which are moſt eligible for the given ſituation.

SECTION THE FOURTH.

THE METHOD OF TRAINING.

MUCH, very much indeed, depends upon this part of the buſineſs: nevertheleſs, it is the common idea of planters of hedges, everywhere, that, having performed the buſineſs of *planting*, and having

having *made* a fence ſufficient to guard the plants, at the time of planting, *their* part is *finiſhed*; the reſt is of courſe left to nature and chance.

THE repairing of the fence;

THE cleaning, &c. of the plants; and the

TRIMMING or pruning them; are not however leſs neceſſary operations, than the planting and fencing; for without proper attention to thoſe, the expence beſtowed upon theſe is only ſo much thrown away. A ſingle gap, eſpecially where ſheep are to be fenced againſt, may cauſe to be undone, in half an hour, what has been doing for two or three years.

IN this point of view, a deep ditch fence is preferable to one raiſed upon the ground; provided the ditch be kept *pointed*; for without this precaution, a ditch, unleſs it be very deep indeed, muſt not be depended upon, as a fence, either againſt cattle or ſheep: but neither the one nor the other will truſt themſelves in a ditch, without a bottom for them to ſtand upon; nothing, indeed, is more terrible to them; eſpecially if part of the mold be formed into a ſharp banklet, placed on the outer brink of the ditch.

HARES are great enemies to young Hedges: a ditch fence is the beſt preſervative againſt them

(paling

(paling or other close fences only excepted). An offset, however, is favorable to them; they will run along it, and crop the plants from end to end: therefore, where hares are numerous, a tufted branch of Furze, Thorns, Holly, or other rough wood, should be stuck, here and there, upon the platform, to prevent their running along it.

THE next business is WEEDING, either with the hoe or by hand; the former is more eligible, where it can be used; as breaking the earth about the roots of the plants is of great service.

FERN is a great enemy to young hedge plants; it is difficult to be drawn by hand, without endangering the plants; and, being tough, it is equally difficult to be cut with the hoe; and, if cut, will presently spring up again. The best manner of getting rid of it, when grown to a head, is to give the stem a twist, near the root,. and let the top remain on, to wither and die, by degrees: this not only prevents its immediate springing; but, to all appearance, destroys the root.

THISTLES, docks, and other tall weeds, are equally injurious to the tender plants, in robbing them of their nourishment, and drawing them up weak and slender, or smothering them out-right, if not timely relieved by the fostering hand of the

 planter.

planter. Even the graſſes are offenſive, and ſhould be extirpated, with all the care and attention neceſſary in a ſeed bed or nurſery.

Nor is it enough to defend the young plants from animal and vegetable intruders; the plants themſelves muſt, by judicious PRUNING, be taught how to grow, ſo as to beſt anſwer the purpoſe for which they are intended.

The Hawthorn is naturally a ſhrubby plant, throwing out ſtrong lateral ſhoots, down cloſe to the ground; more eſpecially when planted by the ſide of a ditch, which, by giving room, favors this propenſity. Theſe horizontal branches, of courſe, draw off their ſhare of nouriſhment from the root; which nouriſhment would be better expended upon the more upright ſhoots. They are, at the ſame time, in the weeder's way, and, by ſtraggling acroſs the ditch, become a temptation to cattle. They ſhould, therefore, from time to time, be ſtruck off with a ſharp inſtrument, either of the hook or the ſabre kind.

In performing this, one rule muſt be obſerved, invariably; that is, to leave the under ſhoots the longeſt, tapering the hedgeling upwards; being very careful, however, not to top the leading ſhoots; for, by doing this, the upward tendency of of the hedge will be checked; and, while its face is kept pruned in the manner here deſcribed, there is no fear of its becoming thin at the bottom.

THUS far we have been ſpeaking of the method of training the SINGLE HEDGE, whether of Crab or Hawthorn. In raiſing the HOLLY, under either of theſe plants, a different kind of pruning is neceſſary: for, notwithſtanding the Holly will ſtruggle, in a ſurprizing manner, under the ſhade and drip of other plants, yet the more air and head room it is allowed, the greater progreſs it will make. In this caſe, therefore, the deciduous plants ſhould be pruned to ſingle ſtems, in the nurſery manner; for all that is required of theſe is ſtrength and tallneſs; the Holly being a ſufficient guard at the bottom.

THIS may be thought an endleſs buſineſs, by thoſe who have not practiſed it; but is it not equally endleſs to prune the young plants of a nurſery? And we here beg leave to remind the young planter, that if he does not pay that care and attention to his hedgelings, in every ſtage of the buſineſs, as he does to his nurſery rows, he is a ſtranger to his own intereſt. The advantage of obtaining a live fence, *on a certainty*, in ſeven or eight years, compared with that of *taking the chance* of one, in fifteen or twenty, is ſcarcely to be done

 away

away by any expence whatever, beſtowed upon planting and training it.

We are, indeed, ſo fully impreſſed with this idea, that we believe every Gentleman would find his account in having even his ſingle Hedges trained with naked ſtems, in order that they might the ſooner arrive at the deſirable ſtate above deſcribed,—*a range of living piles*. We wiſh to be underſtood, however, that we throw this out as a hint to thoſe who wiſh to excel in whatever they undertake, rather than to recommend it, as a practice, to Hedge planters in general.

Nevertheless, we recommend, in general terms, and in the ſtrongeſt manner, to keep the face of a young Hedge pruned, in the manner above deſcribed: or, if the plants be browzed by cattle, or otherwiſe become ſtinted and ſhrubby, to have them cut down, within a few inches of the ground; and by this and every other method promote, as much as poſſible, their upward growth. It is ſome time before a young Hedge becomes an abſolute fence, againſt reſolute ſtock; and the ſhorteſt way of making it a *blind*, is, by encouraging its upward growth, to raiſe it high enough to prevent their looking over it; and, by trimming it on the ſides, to endeavour to render it thick enough, to prevent their ſeeing through it; giving it thereby

thereby the appearance, at leaſt, of a perfect fence.

A HEDGE, pruned with naked ſtems, requires a different treatment, to perfect it as a fence. As ſoon as the ſtems have acquired a ſufficient ſtability, they ſhould be cut off, hedge height; and, in order to give additional ſtiffneſs, as well as to bring the live ſtakes into drill, ſome ſtrong dead ſtakes ſhould be driven in, here and there. This done, the whole ſhould be tightly eddered together, near the top. As an adequate fence againſt horned cattle, the ſtems are required to be of conſiderable thickneſs; but as a ſufficient reſtraint to ſheep only, ſtrong plants may be thus treated, a few years after planting; eſpecially thoſe of the Crab buſh. Upon a ſheep farm, pruning the plants would be eligible, were it only for the purpoſe of getting their heads out of the way of their moſt dangerous enemies.

SECTION THE FIFTH.

MANAGEMENT OF GROWN HEDGE-ROWS.

I. MANAGEMENT of the HEDGE. There is one general rule to be obſerved, in this buſineſs; —*cut often:* for the countryman's maxim is a good one;—" Cut thorns and have thorns."

THE proper length of time *between the cuttings* depends upon the plant, the ſoil, and other circumſtances: eight or ten years may be taken as the medium age, at which the Hawthorn is cut in moſt countries.

IN Norfolk, however, the Hedges are ſeldom cut under twelve to fifteen years; and are ſometimes ſuffered to run twenty and even thirty years, without cutting! The conſequence is, the ſtronger plants have, by that time, arrived at a tree-like ſize, while the underlings are overgrown and ſuffocated *: the number of ſtems are reduced in proportion, and, at that age, it is hazardous to fell the few which remain.

IN Surrey and Kent, ſeven or eight years old is the uſual age at which the Farmers cut down their Quickſet Hedges: and, in Yorkſhire, they are frequently cut ſo young as five or ſix. This may be one reaſon of the excellency of the Yorkſhire Hedges; for, under this courſe of treatment, every ſtem, whether ſtrong or weak, has a fair chance: the weak ones are enabled to withſtand ſo ſhort a ſtruggle, while the large ones are rather invigorated, than checked, by ſuch timely cropping.

* For a remarkable inſtance of this, ſee MID. ECON. Vol. II. p. 383.

WITH

WITH respect to the *first cutting*, this also must be guided by circumstances: a full-stemmed thriving Hedge may stand from twenty to thirty years, between the planting and the first fall; but, if the plants get mossy, or grow shrubby and flat-topped, or put on any other appearance of being diseased or stinted; or, if they are unequal in strength, so that the weaker are in danger of suffering; or, if a young Hedge be much broken into gaps, or any other way rendered defective as a fence, the sooner it is cut down the better; for time will not mend it, and tampering with it will make it worse: whereas cutting it down, within a few inches of the ground, will give a salutary relief to the roots, and the fresh shoots will furnish a full supply of stems; without which no Hedge can be deemed perfect.

THE usual *time of cutting* is during the spring months of February, March, April. The Hawthorn, however, may be cut any time in winter; and it is observable, that the shoots from the stools of Hedges cut in May, when the leaves were breaking forth, have been equally as strong as those from Hedges felled early in the spring. This late felling, however, is not recommended as a practice; the brush wood, cut out at that time, being of less value, than that which is cut when the sap is down.

THE *methods of cutting* are various. In Surrey and Kent, the general practice is to fell to the ground, scour out the ditch, set a stake-and-edder Hedge behind or partially upon the stubs, and throw some rough thorns into the ditch.

IN Norfolk, there are two ways practised: one, to cut within a few inches of the face of the bank, remake the ditch and bank, and set a brush hedge as for the original planting: the other is called *Buckstalling:* which is to leave stems, about two feet long,—without repairing the bank or setting a Hedge; and only shovelling out the best of the mold of the ditch, to form the bottoms of dung-hills with. This is a much cheaper way than the other, and where the Hedge stands at the foot of the bank, and remains full stocked with stems, it is not ineligible; especially if a few of the slenderest of the old shoots be layered in, between the bank and the stems, and kept there by a coping sod, taken from the foot of the back of the bank: but when the roots lie high in the bank, and are of course more or less exposed by the soil's moldering away from them into the ditch, such treatment is destructive to the Hedge; which, in this case, requires to be cut down, within a few inches of the roots, every eight or ten years, the ditch to be scoured, and the bank to be faced and made fencible, by a brush Hedge. This circumstance, alone,

alone, furniſhes ſufficient argument againſt planting high in the bank.

In Hertfordſhire, Glouceſterſhire, and ſome parts of Yorkſhire, *Plaſhing* is much in uſe. This is done by cutting the larger ſtems, down to the ſtub, and topping thoſe of a middling ſize, hedge height, by way of ſtakes, between which the moſt ſlender are interwoven, in the wattle manner, to fill up the interſtices, and give an immediate live fence. If live ſtakes cannot be had, dead ones are uſually driven in their ſtead: and, in order to keep the plaſhers in their places, as well as to bring the ſtakes into a line and ſtiffen the whole, it is cuſtomary, in moſt places, to edder ſuch Hedges.

If the ſtems, alone, are not ſufficient as a fence, this method of treatment may, in ſome caſes, be eligible; provided it be properly executed: much, however, depends upon the manner of doing it: many good Hedges have been ſpoiled by plaſhing. The plaſhers ſhould be numerous, and ſhould be trimmed to naked rods, in order that their ſpray may not incommode the tender ſhoots from the ſtools below *: they ſhould be laid in an aſcending direction,

* The moſt effectual method of preventing this evil,—as well as that ariſing from live ſtakes throwing out buſhy tops, to the injury of the ſhoots from the lower ſtubs,—is to drive a line of dead ſtakes, a few

direction, so that they may be bent without nicking at the root, if possible: such as will not stoop, without danger of breaking, should be nicked with an *upward*, not with a *downward* stroke: *that*, if properly done, gives a *tongue* which conducts the rain water from the wound; *this*, a *mouth* to catch it.

HOWEVER, in cases where the *stems* stand regular, and are, in themselves, stiff enough for a Fence, or where they can be readily made so, by driving large stakes in the vacancies and weak places, plashing, and every other expedient, ought to be dispensed with:—where, upon examination, the stems are found insufficient, it is generally the best practice to fell the whole to the ground, and train a set of new ones.

IN case of gaps or vacancies too wide to be filled up by the natural branches of the contiguous stools, they should be filled up by *layering* the neighbouring young shoots, the first or second year after felling; being careful to weed and nurse up the young layers, until they be out of harm's way. If such vacancies be numerous, it is best to keep

a few inches from the line of stubs, and to wind the plashers among these stakes; thus leaving the young shoots a free air to rise in, and at the same time, forming a live hedge to protect them. See more of this method of plashing, in MID. ECON. VOL. I. p. 92.

the

the whole Hedge, let its ſituation be what it may, trimmed low, in order to give air and headroom to the layers.

ALL fallen Hedges, whether layered or not, ſhould be *pruned* on the ſides, the firſt and ſecond years after felling; at the ſame time *weeding* out the brambles, thiſtles, docks, and every other *weed*, whether herbaceous or ligneous; which, by crouding the bottom, prevent the young branches from uniting, and interweaving with each other.

THE proper time for performing this, is when the thiſtles are breaking into blow, before their ſeeds have acquired a vegetative body. The large Spear Thiſtle (*Carduus Lanceolatus*), ſo miſchievous in young Hedges, and ſo conſpicuouſly reproachful to the Farmer, when its ſeeds are ſuffered to be blown about the country, is a biennial plant, which does not blow till the ſecond year; when, having produced its ſeed, the root dies: it is therefore unpardonable to neglect taking this in the criſis; for, by ſo doing, the whole race becomes at once extirpated.

THE fitteſt inſtrument for the purpoſe of ſtriking off the ſide ſhoots, and for weeding, is a long hook, or rather a long ſtrait blade with a hooked point, which is convenient for cutting out the brambles

and

and weeds, that grow in the middle of the Hedge, as well as for other purpoſes. We will venture to ſay, that whoever puts this piece of huſbandry in practice once, will not neglect doing it a ſecond time; the uſes, as well as the neatneſs, reſulting from it are numerous, and the expence of performing it little or nothing.

IF the Hedge be intended to run up, either as a ſource of uſeful materials, or as a ſhelter in graſs-land incloſures, the leading ſhoots ſhould not be touched; nevertheleſs, it ought, in theſe two early trimmings, to be kept thin towards the top, leaving it to ſwell out thicker towards the bottom: but if it be intended to be kept down, as we have already ſaid it ought to be, between arable incloſures, the leading ſhoots ſhould be cropped low, both the firſt and the ſecond year; in order to check its upward tendency, and give it a dwarfiſh habit; and the cropping muſt be repeated, from time to time, as occaſion may require.

A HEDGE under this treatment becomes a perpetual Fence, and its duration might be deemed everlaſting. The age of the Hawthorn is probably unknown; but ſuppoſing that it will bear to be felled every ten years, for two hundred years, during which time there will be twenty falls of wood (what a mountainous pile for one ſlip of land and

one

one ſet of roots to produce!) may we not be allowed to ſuppoſe, that a ſimilar hedge, kept in a dwarfiſh ſtate (in which ſtate its produce, and conſequently its exhauſtion, could not be one tenth ſo much as that in the former ſuppoſition) would live to the age of three or four hundred years? Tenants have only a temporary property in the hedges of their reſpective farms; and it is the buſineſs of landlords, or their agents, to ſee that they are properly treated. The value of an eſtate is heightened, or depreciated, by the good or bad ſtate of its fences; which, it is well known, are expenſive to raiſe, and, when once let down, are difficult to get up again.

WITH reſpect to the *rough* and the *worn-out Hedges*, which conſtitute a large majority of the Hedges of this country, it is not an eaſy matter to lay down any preciſe rules of treatment. If the ground they grow in be ſufficiently moiſt, they may be helped by felling, and layering, in the manner already deſcribed, or by filling up the vacancies with young quicks, or with the cuttings of ſallow, elder, &c. &c. firſt clearing the ground from ivy, and other encumbrances; but, in a dry bank, which has been occupied by the roots of trees and ſhrubs for ages, and which, by its ſituation, throws off the rain water that falls upon it, there can be little hope either of plants or cuttings taking to advantage.

THE

THE beſt aſſiſtance that can be given, in this caſe, is to drive ſtakes into the vacancies, and interweave the neighbouring boughs between the ſtakes, training them in the eſpalier manner: or, if the vacancies be wide, to plaſh tall boughs into them.

THESE, however, are only temporary reliefs; for, if the bodies of the plants themſelves be ſuffered to run up, and to draw the nouriſhment from the plaſhers, the breaches will ſoon be opened again, and it will be found difficult to fill them up a ſecond time: the only way by which to render this method of treatment in any degree laſting, is, to keep the whole hedge trimmed, as ſnug and low as the purpoſe for which it is intended will permit; weeding it with the ſame care as a young Hedge. By this means the vacancies in time will grow up, and one regularly interwoven ſurface will be formed.

AFTER all, however, an old worn-out Hedge, with all the care and attention that can be beſtowed upon it, cannot continue for any length of time; and whenever it verges upon the laſt ſtage of decline, it is generally the beſt management to grub it up at once, and raiſe a new one in its place; otherwiſe the occupier muſt be driven, in the end, to the humiliating and diſgraceful

graceful expedient of patching with dead Hedge-work.

We are happy in having it in our power to ſay, that the practice of *replanting* Hedges has, of late years, become prevalent in a county which has long taken the lead in many important departments of huſbandry; and, although we have had occaſion to cenſure ſome of its practices, with reſpect to Fences, we have great pleaſure in giving to it due praiſe, in this particular; we ſpeak of the county of Norfolk. The beſt way is to level the old bank, about michaelmas, in order that the mold may be thoroughly moiſtened by the winter's rains, and tempered by the froſts. The roots and old ſtems will, in general, more than repay the expence of grubbing and levelling, and when the old ſtools are numerous, and fuel is dear, will, ſometimes, go a good way towards raiſing the new Fence. One great advantage, ariſing from this practice, in an arable country, is doing away the crookedneſs of old Hedges.

There is one general rule to be obſerved, in renewing a Hedge in this manner, which is to plant a ſpecies of Hedgewood different from that which formerly occupied the ſoil; and we know of no better change, after the Hawthorn, than the Crab tree and Holly.

II. MANAGE-

II. MANAGEMENT OF HEDGEROW TIMBER.—Thus, having mentioned the ſeveral ways of raiſing and repairing LIVE HEDGES *, we now come to the training, and general treatment, of HEDGEROW TIMBER: and, firſt, as to the young Oaks, which we recommended to be planted with the Hedgewood.

THE moſt eligible length of ſtem has been mentioned to be from fifteen to twentyfive feet; and, with due attention to their leading ſhoots, there will be little difficulty in training them, on a good ſoil, to that or a greater height. If, by accident or diſeaſe, the head be loſt, the ſtem ſhould be taken off at the ſtub, and a freſh ſhoot trained. However, in this caſe, if the Hedge be got to any conſiderable height, it is beſt to let the ſtump ſtand, until the firſt fall of the Hedgewood; for, then, the young tree may be trained with leſs difficulty.

NEXT to the danger of the young trees being cropt by cattle, is that of their being hurt by the Hedgewood: firſt, from their being overhung and

* For farther remarks on this ſubject, ſee YORK. ECON. Art. FENCE.; and MID. ECON. under the ſame Title.

ſmothered

ſmothered amongſt it; ſecondly, from their being drawn up too tall and ſlender; thirdly, from their being chaſed againſt the boughs by the wind; and, laſtly, from their ſtems getting locked in between the branches, ſo as to cauſe an indenture in the ſtem, and thereby render it liable to be broken off by the wind. The ſimpleſt way of guarding againſt theſe evils, is, to keep the Hedgewood down to fence height; otherwiſe, great care and attention are requiſite in training Hedge timber. Even in this caſe, the plants ſhould be frequently looked over,—to ſee that the lower parts of them do not interfere with the ſtems of the Hedgewood, —to take off, as occaſion may require, the lateral ſhoots,—and to give ſimplicity and ſtrength to the leaders, until the plants have acquired a ſufficient length of ſtem.

WHEN this is obtained, it may not be amiſs to endeavour to throw the general tendency of the head to one or the other ſide of the Hedge, in order to give air and head room to the plants, and *crookedneſs* to the timber. In ſhort, if trees in Hedges are not treated with the ſame attention as thoſe in Nurſeries and Plantations, it were better not to plant them; as they will become an encumbrance to the Hedge, without affording either pleaſure, or profit, to the planter, or his ſucceſſors.

WHAT remain now to be considered are, the GROWN TIMBERS, the TIMBER STANDS, and the POLLARDS with which old Hedges are frequently stored.

THERE is not a more absurd practice, in the circle of rural affairs, than that of making FALLS of HEDGEROW TIMBER; which is neither more nor less than for the woodman to begin at one end of the Hedge, and hack down every timber tree he comes at, whether full-grown, over-grown, or only half-grown, until he reaches the other. The impropriety is the same, whether a young thriving tree be taken down before it has arrived at its full growth, or an old one be suffered to remain standing, after it has entered upon the stage of decline.

A TIMBERED estate should frequently be gone over, by some person of judgement; who, let the price and demand for timber be what they may, ought to mark every tree which wears the appearance of decay. If the demand be brisk, and the price high, he ought to go two steps farther, and mark not only such as are full grown, but such, also, as are near perfection; for the interest of the money, the disencumbrance of the Hedge and the neighbouring young timbers, and the comparative advantages of a good market, are not to be bartered for any increase of timber, which can reasonably

reaſonably be expected from trees in the laſt ſtage of their growth.

THERE are men in this kingdom, who, from miſmanagement of their timber, are now loſing, annually, very handſome incomes. The loſs of price which generally follows the refuſal of a high offer, the certain loſs of intereſt, the decay of timber, and the injuries ariſing from the encumbrance of full-grown trees, are irretrievable loſſes, which thoſe who have the care and management of timber ſhould ſtudiouſly endeavour to avoid.

BUT while we thus hold out the diſadvantages of ſuffering timber to ſtand until it be overgrown, it is far from our intention to recommend, or even countenance, a premature felling,—of Hedgerow timbers more particularly: for although, in woods and cloſe groves, a ſucceeding crop of ſaplings may repair, in ſome degree, the loſs of growth, in timber untimely fallen; yet it is not ſo in Hedges,—where ſapling ſtands are liable to be ſplit off from the ſtool, as ſoon as they acquire any conſiderable top; as being expoſed ſingly, and on every ſide, to the wind: and all that can be expected from the ſtools of trees in Hedgerows, is a ſufficiency of ſhoots to fill up the breaches in the Hedge.

WITH respect to POLLARDS * in Hedges, some general rules are observable. Pollards, which are fully grown, but yet remain sound, should be taken down, before they become tainted at the heart;—for a good gate post is worth five shillings; but a firing Pollard, of the same size, is not worth one shilling. Firing Pollards which, by reason of their decay, or stintedness, will not, in the course of eighteen or twenty years, throw out tops equal in value to their present bodies, should also be taken down;—for the principal and interest of the money will be worth more, at the end of that time, than the body and top of the Pollard; besides the desirable riddance of such unsightly encumbrances. But, in case a Pollard is already so much tainted as to be rendered useless as timber, yet sound enough, to all present appearances, to throw out, in the time abovementioned, a top or tops of more than equal value to its present body;—it rests upon a variety of circumstances, whether, in strict propriety of management, such Pollard ought to stand or fall.

WE declare ourselves enemies to Pollards; they are unsightly; they encumber and destroy the Hedge they stand in (especially those whose stems

* Trees which have been *polled,* topt, or headed down to the stem.

are

are ſhort), and occupy ſpaces which might, in general, be better filled by timber trees; and, at preſent, it ſeems to be the prevailing faſhion to clear them away: nevertheleſs, in a country, in which woodlands and coppices are ſcarce, Hedge pollards furniſh a valuable ſupply of fuel, ſtakes, &c.—and every man who clears away the claſs of Pollards laſt-mentioned, without planting an adequate quantity of coppice wood, commits a crime againſt poſterity; more eſpecially in a diſtrict which depends wholly upon the ſea for a ſupply of coals. For, although Great Britain is, at preſent, miſtreſs of her own coaſt, what man is raſh enough to ſay, that, amidſt the revolutions in human affairs, ſhe will always remain ſo?

WITH reſpect to the YOUNG TIMBERS, which frequently abound in rough Hedgerows, we venture to recommend the following management.

UPON eſtates whoſe Hedge timber has been little attended to (and, we are ſorry to ſay, ſuch are nine tenths of the eſtates in the kingdom), the firſt ſtep is to ſet out the plants, and clear away the encumbrances.

AFTER what has been ſaid, it may be needleſs to repeat, here, that, where the choice reſts upon the ſpecies of tree, the Oak ſhould invariably be chosen; for every other species we consider as a kind of encumbrance, which ought to be done away, as soon as it can with any colour of propriety.

It is bad practice to permit Hedges to remain crouded with timber stands; they should, in general, be set out singly, and at distances proportioned to their respective sizes; so that their tops be not suffered to interfere too much with each other.

There is, however, one exception to this rule: where two trees, standing near each other, have grown up, in such a manner, that their joint branches form, in appearance, but one top, they should both be permitted to stand; for if one of them be removed, the other will not only take an unsightly outline, but will receive a check in its growth, which it will not overcome for several years. It is, nevertheless, observable, that twin trees, as well as those which are double-stemmed, are dangerous to stock: not only cattle, but even horses, have been known to be strangled, by getting their heads locked in between them.

The method of training the young plants has already been described; it now only remains to say a few words, as to the Pruning and Setting up Hedgerow timbers.

Low-

LOW-HEADED trees have been already condemned, as being injurious to the Hedge, as well as to the Corn which grows under them. To remove or alleviate theſe evils, without injuring the tree itſelf, requires the beſt ſkill of the woodman. The uſual method is to hack off the offending bough; no matter how nor where; but, moſt probably, a few inches from the body of the tree, with an axe; leaving the end of the ſtump ragged, and full of clifts and fiſſures, which, by receiving and retaining the wet that drips upon them, render the wound incurable. The mortification, in a ſhort time, is communicated to the ſtem, in which a receſs or hollow being once formed, ſo as to receive and retain water, the decline of the tree, though otherwiſe in its prime, from that time, muſt be dated; and, if not preſently taken down, its properties, as a timber tree, will, in a few years, be changed into thoſe of firewood only. How many thouſand timber trees ſtand, at this hour, in the predicament here deſcribed; merely through injudicious lopping! It is this improper treatment, which has brought Hedgerow timber into a diſrepute otherwiſe undeſerved.

THERE is a wonderful ſimilarity in the operations of Nature upon the Vegetable and the Animal Creation. A ſlight wound in the Animal Body ſoon heals up, and ſkins over, while the wound ſucceeding the amputation of a limb, is with difficulty cicatrized. The effects are similar with respect to the Vegetable Body: a twig may be taken off with safety, while the amputation of a large bough will endanger the life of the tree. Again, pare off a small portion of the outer bark of a young thriving tree, the first summer's sap will heal up the wound: if a small twig had been taken off with this patch of bark, the effect would have been nearly the same; the wound would have been cicatrized, or barked over, in a similar manner; and the body of the tree as safely secured from outward injury, as if no such amputation had taken place. Even a considerable branch may be taken off, in this manner, with impunity, provided the surface of the wound be left smooth and flush with the *inner* bark of the Tree; for, in a few years, it will be completely closed up, and secured from injury; though an eschar may remain for some years longer. But if a large bough be thus severed, the wound is left so wide, that it requires, in most trees, a length of time to bark it over; during which time, the body of the tree having increased in size, the parts immediately round the wound become turgid, while the face of the wound itself is thrown back into a recess; and, whenever this becomes deep enough to hold water, from that time the wound is rendered incurable: Nature has, at least, done her part; and, whether or not, in this case,

caſe, aſſiſtance may be given, by opening the lower lip of the wound, remains yet (it is probable) to be tried by experiment: until that be aſcertained, or ſome other certain method of cure be known, it were the height of imprudence to riſk the welfare of a Tree on ſuch hazardous treatment.

FURTHER, although a branch of conſiderable ſize may be taken off, cloſe to the body of the Tree, with ſafety; yet, if the ſame branch be cut *a few inches* from it, the effect is not the ſame; for, in this caſe, the ſtump generally dies; conſequently, the cicatrization cannot take place, until the ſtem of the Tree has ſwelled over the ſtump, or the ſtump has rotted away to the ſtem; and, either way, a mortification is the probable conſequence. Even ſuppoſing the ſtump to live, either by means of ſome twig being left upon it, or from freſh ſhoots thrown out, the cicatrization, in this caſe, will be ſlow (depending entirely upon the feeble efforts of the bark of the ſtump); and before it can be accompliſhed, the Tree itſelf may be in danger. But, had the amputation been made *at a diſtance* from the ſtem, and immediately *above a twig*, ſtrong enough to draw up a ſupply of ſap, and keep the ſtump alive, with certainty, no riſk would have been

been incurred; eſpecially if the end of the ſtump had been left ſmooth, with the ſlope on the under ſide, ſo that no water could hang, nor receſs be formed.

FROM what has been ſaid, the following general rules, with reſpect to ſetting up low-headed trees, may, we humbly conceive, be drawn with ſafety: *ſmall boughs ſhould be cut off, cloſe to the ſtem; but large ones, at a diſtance from it; and above a lateral branch, large enough to keep the ſtump alive.* Thus, ſuppoſing the ſtem of a tree, in full growth, to be the ſize of a man's waiſt, a bough the thickneſs of his wriſt may be taken off, with ſafety, near the ſtem; but one as thick as his thigh ſhould be cut at the diſtance of two feet from it, at leaſt: leaving a ſide branch, at leaſt an inch in diameter, with a top in proportion, and with air and head-room enough to keep it in a flouriſhing ſtate. For this purpoſe, as well as for the general purpoſe of throwing light into the head, the ſtanding boughs ſhould be cleared from their lower branches, particularly ſuch as grow in a drooping direction. In doing this, no great caution is required; for, *in taking a bough from a bough*, let their ſizes be what they may, little riſque can be thereby incurred, upon the *main body of the tree*.

THERE

THERE is another general rule, with regard to pruning trees. The bough ſhould be taken off, either by the *upward ſtroke* of a ſharp inſtrument (and, generally ſpeaking, *at one blow)*, or with a ſaw: in the latter caſe, it ſhould previouſly be notched, on the under ſide, to prevent its ſplitting off, in the fall. If the bough to be taken off be heavy, the ſafeſt way is, firſt to cut it off, a few inches from the ſtem, with an axe, and then to clear away the ſtump, cloſe and level, with a ſaw; doing away the roughneſſes, left by the teeth of the ſaw, with a plane, or with a broad-mouthed chiſſel, or an axe; in order to prevent the wet from hanging in the wound. A ſaw, for this purpoſe, ſhould be ſet very wide; otherwiſe, it will not make its way through green wood.

THE fitteſt opportunity for pruning and ſetting up young timbers, as well as for taking down Pollards and dotard timbers, and clearing away other encumbrances, is when the Hedge itſelf is felled; and it were well for landed individuals (as for the Nation at large) if no Hedge was ſuffered to be cut down, without the whole buſineſs of the Hedgerow being, at the ſame time, properly executed.

FOR

FOR farther Information respecting HEDGES and HEDGEROW TIMBER, see the RURAL ECONOMY of YORKSHIRE, Vol. I. p. 201.

ALSO the RURAL ECONOMY of the MIDLAND COUNTIES, Vol. I. pages 83 to 95, and the MINUTES thence referred to.

DIVISION

DIVISION THE FOURTH,

WOODLANDS; OR, USEFUL PLANTATIONS.

INTRODUCTORY REMARKS.

ALTHOUGH it may be difficult to diſtinguiſh, preciſely, between *uſeful* and *ornamental* plantations, yet the diſtinction between a rough coppice, in a recluſe corner of an eſtate, and a flowering ſhrubery, under the windows of a manſion, is obvious: the one we view as an object of *pleaſure* and amuſement, while the other is looked upon in the light of *profit*, only. Upon theſe premiſes we ground our diſtinction. Under the preſent head, we purpoſe to ſpeak of plantations, whoſe leading features are of the more uſeful kind, and whoſe principal end is profit; reſerving thoſe, whoſe diſtinguiſhing characteriſtics are ornamental,

mental, and whose primary object is pleasure, for the SECOND PART of this VOLUME.

PERHAPS, it will be expected, that, before we begin to treat of the propagation of TIMBER, we should previously prove an approaching SCARCITY of that necessary article in this country: for it may be argued, that every acre of land applied to the purposes of planting, is lost to those of agriculture; and, as far as *culturable* land goes, the argument is just. To speak of this subject, generally, as to the whole kingdom, and at the same time precisely, is perhaps what no man is prepared for.

FROM an extensive knowledge of the different parts of the kingdom, we believe that the Nation has not, yet, experienced any real want of timber. We are happy to find, that, in many parts of it, there are great quantities now standing; while, in many other parts, we are sorry to see an almost total nakedness. With respect to large well grown OAK TIMBER, such as is fit for the purposes of SHIP BUILDING, we believe there is a growing scarcity, throughout the kingdom.

WE will explain ourselves, by speaking particularly as to one district—the VALE OF PICKERING, in Yorkshire. This district, for ages past, has supplied,

ſupplied, in a great meaſure, the ports of Whitby and Scarborough with ſhip timber. At preſent, notwithſtanding the extenſive tracts of Woodlands ſtill remaining, there is ſcarcely a tree left ſtanding with a load of timber in it. Beſides, the woods which now exiſt, have principally been raiſed from the ſtools of timber trees, formerly taken down; the ſaplings from which being numerous, they have drawn each other up ſlender, in the grove manner; and, conſequently, never will be ſuitable to the more valuable purpoſes of the ſhip builder.

WHEN we conſider the prodigious quantity of timber which is conſumed in the conſtruction of a large veſſel, we feel a concern for the probable ſituation of this country, at ſome future period. A SEVENTY-FOUR GUN SHIP (we ſpeak from *good authority)* ſwallows up three thouſand loads of Oak timber. A load of timber is fifty cubical feet; a ton, forty feet; conſequently, a ſeventy-four gun ſhip takes 2,000 large well grown timber trees; namely, trees of nearly two tons each!

THE diſtance recommended, by authors, for planting trees, in a *Wood*, (a ſubject we ſhall ſpeak to particularly in the courſe of this chapter) in which Underwood is alſo propagated, is thirty feet or upwards. Suppoſing trees to ſtand at two rods (33 feet, the diſtance we recommend they

ſhould

ſhould ſtand at, in ſuch a plantation), each ſtatute acre would contain 40 trees; conſequently, the building of a ſeventy-four gun ſhip would clear, of ſuch Woodland, the timber of 50 acres. Even ſuppoſing the trees to ſtand at one rod apart (a ſhort diſtance for trees of the magnitude above-mentioned), ſhe would clear twelve acres and a half; no inconſiderable plot of Woodland. When we conſider the number of king's ſhips that have been built during the late wars, and the Eaſt Indiamen, merchants ſhips, colliers, and ſmall craft, that are launched daily in the different ports of the kingdom, we are ready to tremble for the conſequences. Nevertheleſs, there are men who treat the idea of an approaching ſcarcity as being chimerical; and, at preſent, we will *hope* that they have ſome foundation for their opinion, and that the day of want is not near. At ſome future opportunity, we may endeavour to reduce to a degree of certainty, what at preſent is, in ſome meaſure, conjectural. The preſent ſtate of this iſland with reſpect to ſhip timber is, to the community, a ſubject of the very firſt importance.

HOWEVER, in a work like the preſent, addreſſed to individuals, rather than to the nation at large, a true eſtimate of the general plenty or ſcarcity of timber is only important, as being inſtrumental in aſcertaining the local plenty, or ſcarcity, which is likely

likely to take place in the particular neighbourhood of the planter. This may be called a new doctrine, in a Treatise on Planting. It is so, we believe, and we wish to have it understood, that we address ourselves to the PRIVATE INTEREST, rather than to the public spirit, of our readers; and we appeal to every one, who has had extensive dealings with mankind, for the propriety of our conduct.

WE are well aware that, situated as this country appears to us to be at present, Planting ranks among the first of public virtues; nevertheless, we rather wish to hold out that *lasting fame*, which always falls to the share of the successful planter, and those *pecuniary advantages*, which must ever result from plantations, judiciously set about and attentively executed, as being motives of a more *practical* nature.

WE wish, in the first place, to do away a mistaken notion, that when once a piece of ground is set apart for a plantation, it becomes a dead weight upon the estate, or a blank in it, at least. Nothing can be less true; for plantations, entered upon with judgement, and carried on with spirit, accumulate in value, as money at interest upon interest. If an estate, after a plantation has been made upon it, is not worth more, by the trouble

 and

and expence of making it, than it was before, the undertaking was either ill judged, or badly executed.

An Ozier bed riſes to profit the ſecond or third year, and a Coppice in fifteen or twenty; while an Oak may be a century before he reach the moſt profitable ſtate: but do they not, in effect, all pay an annual income? Do not eſtates ſell at a price proportioned to the value of the timber which is upon them? and does not this value increaſe annually? The ſweets of a fall are well underſtood, and the nearer we approach to this, the more valuable are the trees to be fallen.

We have ſome knowledge of a Gentleman, now living, who, during his lifetime, has made plantations, which, in all probability, will be worth, to his ſon, as much as the reſt of his eſtate; handſome as it is. Suppoſing that thoſe plantations have been made fifty or ſixty years, and that, in the courſe of twenty or thirty more, they will be worth 50,000l. may we not ſay that, at preſent, they are worth ſome twenty or thirty thouſand? What an incitement to planting!

Every thing, however, depends upon management. It is not ſticking in a thouſand or ten thouſand plants, as if for the ſole purpoſe of ſaying, "I have done thoſe things," without giving them a ſecond a ſecond thought, that will ever bring in the profits of planting; yet, how many Gentlemen do we ſee ſquandering their money, laying their lands waſte, and rendering themſelves ridiculous, by ſuch management!

THE firſt PRECAUTION requiſite to be taken, by a man who wiſhes to ſerve his family and his country, and, at the ſame time, to afford amuſement and acquire credit to himſelf, by planting, is to conſider well his own particular ſituation.

MUCH depends upon *ſoil*, and much on *locality*, or relative ſituation, with reſpect to water carriage, and a variety of other circumſtances; as contiguity to a large town, or a manufacturing place, which generally enhances the value of land, and the price of labour.

MUCH, alſo, depends upon the *natural features*, or poſitive ſituation of his eſtate: the hang of a hill, which is too ſteep for the plow, and a ſwampy bottom, too rotten to bear paſturing ſtock, and which cannot be rendered firm enough for that purpoſe, but at too large an expence, may, in general, be highly improved, by planting *.

* The laſt, however, is a caſe that will now ſeldom occur, ſince the art of DRAINING is ſo well underſtood.

AGAIN, where the top ſoil, or culturable ſtratum, is of an unproductive nature, while a bed of clay, loam, or other good ſoil, lies under it, planting may ſometimes be made greatly advantageous. An inſtance occurs, in the Vale of Glouceſter, of a coppice which pays at the rate of fourteen or fifteen ſhillings an acre, annually; while the land, which ſurrounds it, is not worth more than eight or ten ſhillings. The ſoil is a *ſour* clay, and the ſubſtratum a calcareous loam. The valuable plantations above-mentioned afford a ſimilar inſtance; the top ſoil is a light unproductive ſand, under which lies a thick ſtratum of ſtrong clayey loam. Wherever we ſee the Hawthorn flouriſh upon *bad land*, we may venture to conclude, that, under ordinary circumſtances, ſuch land will pay for planting.

BUT, with reſpect to low lands, which wear a profitable ſward, and will bear the tread of cattle, or which, by judicious draining, can be rendered ſuch, at a reaſonable expence; alſo to uplands, which, by proper management, will throw out profitable crops of corn, and other arable produce, more eſpecially if the ſubſtratum is of a nature ungenial to the ligneous tribes; we are of opinion, that planting can ſeldom be carried on, upon a large ſcale, with propriety. Neverthelcſs, even under theſe circumſtances, ſkreen plantations, upon

upon expoſed heights, as well as ſheltering Groves, and ſtripes or patches of planting, to fill up the inconvenient crookedneſſes of the borders of arable fields, may be productive of real and ſubſtantial improvement to an eſtate.

THE next ſtep, which a Gentleman ought to take, before he ſet about raiſing plantations, upon a large ſcale, is to look round his neighbourhood, and make himſelf acquainted with its preſent ſtate, as to Woodlands; as well as with the comparative value which theſe bear to arable and graſs lands. He muſt go ſtill farther; he muſt learn the natural conſumption of the country; not only of timber in general, but of the ſeveral ſpecies. Nor muſt he ſtop here; he muſt endeavour to pry into futurity, and form ſome judgement of the particular ſpecies, whether it be Oak, Aſh, Elm, Beech, the Aquatics, Pines, or Coppice Woods, which will be wanted, at the time his plantations arrive at maturity.

IT is poſſible, there may be ſituations, in this iſland, where, from a ſuperabundance of Woodlands, it would be unprofitable to plant, even hangs, and bad top ſoils: it is not probable, however, that any ſuch places are to be found; for, in a country ſituated near water carriage, (and if the preſent ſpirit of cutting canals continue to prevail, what part of this island will, a century hence, be out of the reach of water carriage?) ship timber will, in all human probability, always find a market; and, in situations remote from such cheap conveyance, foreign timber will always bear a price proportionably high; consequently, the timber raised, in such a country, will, in all probability, find a market in the neighbourhood of its growth.

BEFORE we begin to speak of the several species of Plantations or Woodlands, and the methods of raising them, it will be proper to ENUMERATE, here, the different SPECIES OF TREES, which we conceive to be most eligible to be planted, for the purposes of timber and underwood, in this country.

UNDER the article CHOICE OF TIMBER TREES, it appears that

THE OAK,
THE ASH,
THE ELM, and
THE BEECH,

are the four principal *domestic* timbers, now in use, in this kingdom: To which must be added

THE PINE TRIBE, particularly
THE LARCH; and
THE AQUATICS;

as

as ſubſtitutes for *foreign* timber, at preſent imported, in vaſt quantities, into this iſland: And to thoſe muſt be added, as *coppice woods*;

THE ASH,
THE CHESNUT,
THE HAZEL,
THE SALLOW,
THE WILD SORB, and
THE OZIER *.

THERE are four diſtinct SPECIES of WOODLANDS:

WOODS,
TIMBER GROVES,
COPPICES,
WOODY WASTES.

BY a *Wood* is meant a mixture of timber trees and underwood; by *Timber Grove*, a collection of timber trees only, placed in cloſe order; by *Coppice*, ſtubwood alone, without an intermixture of timber trees; and by *Woody Waſte*, graſs land over-run with rough woodineſs; or a mixture of Woodland and graſſy patches; which being thought an object of paſturage, the wood is kept under, by being browſed upon by ſtock, while the graſs, in

* The mode of PROPAGATION, and the SOIL ſuitable to the ſeveral ſpecies, appear under their reſpective names, in the ALPHABET OF PLANTS.

its turn, is ſtinted by the trees, and rendered of an inferior quality, by the want of a free admiſſion of ſun and air.

In practice, theſe Woody Waſtes ought firſt to be taken under conſideration; for while a Gentleman has an acre of ſuch land upon his eſtate, he ought not (generally ſpeaking) to think of ſetting about raiſing original plantations: for, if graſſineſs prevail, and the ſoil be unkind for Wood, let this be cleared away, and the whole be converted to paſture or arable. But if, on the contrary, woodineſs prevails, fence out the ſtock, and fill up the vacancies, in the manner *hereafter* deſcribed: for, in a ſyſtematic Treatiſe upon Planting, we think it moſt conſiſtent with method, to treat of Woodlands in the order already ſet down.

SECTION THE FIRST.

WOODS.

OPEN WOODS are adapted, more particularly, to the purpoſe of raiſing TIMBER for SHIPBUILDING, and, perhaps, for ſome few other purpoſes, where

crookedneſs

crookedness is required. Where a *straightness* and length of stem, and cleanness of grain, are wanted, CLOSE WOODS OR GROVES are more eligible; and, where Stubwood is the principal object, COPPICES, unencumbered with timber trees, are most adviseable.

IT follows, that no timber tree whatever, but the Oak, can be raised, with propriety, in open Woods, and this, only, when a supply of ship timber is intended; consequently, open Woods are peculiarly adapted to places lying conveniently for water carriage, or which may, in all probability, lie convenient for water carriage, a century or two hence.

VARIOUS opinions prevail, with respect to the most eligible METHOD OF RAISING A WOOD: some are warm advocates for *sowing*, others for *planting*; some again are partial to *rows*, while others prefer the *irregular* culture.

THE dispute about sowing and planting may, in some measure, be reconciled in the following manner: Where the strength of the land lies in the substratum, while the surface soil is of an ungenial nature, *sow*, in order that the roots may strike deep, and thereby reap the full advantage of the treasures below: but, on the contrary, when the

top

top ſoil is good, and the bottom of an oppoſite quality, *plant*, and thereby give the roots the full enjoyment of the productive part of the ſoil; or, under theſe laſt circumſtances, *ſow*, and *tap* the young plants as they ſtand (with a tapping inſtrument), and thereby check their downward tendency, as well as ſtrengthen their horizontal roots.

By *this* method of treating ſeedling plants, the peculiar advantage of planting is obtained. The diſpute, therefore, ſeems to reſt entirely upon this queſtion: Which of the two methods is leaſt expenſive? To come at this, there are two things to be conſidered—the *actual expence* of labour and other contingent matters, and the *loſs of time* in the land occupied. With reſpect to the former, ſowing is beyond compariſon the cheapeſt method; but, in regard to the latter, planting may ſeem to gain a preference; for the ſeed bed is ſmall, compared with the ground to be planted, and while that is rearing the ſeedling plants, this continues to be applied to the purpoſes of huſbandry. However, if we conſider the check which plants in general receive in tranſplantation *, and if (as we ſhall hereafter

* We have known an inſtance of tranſplanted Oaks remaining upon the ground ſo long as eight years before they began to move. And let us hear what Miller ſays upon this ſubject; we have no reaſon to doubt his ſpeaking from his own experience,

hereafter ſhew) the interſpaces of an infant Wood may, for ſeveral years after ſowing, be ſtill cultivated to advantage, the preference, we conceive, is evidently, and beyond all diſpute, on the ſide of ſowing.

WITH reſpect to the arrangement of Wood Plants,—the preference to be given to the *row*, or the *random* culture, reſts in ſome meaſure upon the nature and ſituation of the land to be ſtocked with plants. Againſt ſteep hangs, where the plow cannot be conveniently uſed in cleaning and cultivating the interſpaces, during the infancy of the Wood, either method may be adopted; and if plants are to be put in, the *quincunx* manner will be found

rience, though he does not particularize it.—"When Oak trees are cultivated with a view to profit, acorns ſhould be ſown, where the trees are deſigned to grow; for thoſe which are tranſplanted will never arrive to the ſize of thoſe which ſtand where they are ſown, nor will they laſt near ſo long. For in ſome places where theſe high trees have been tranſplanted, with the greateſt care, they have grown very faſt for ſeveral years after, yet are now decaying, when thoſe which remain in the places where they came up from the acorns, are ſtill very thriving, and have not the leaſt ſign of decay. Therefore, whoever deſigns to cultivate theſe trees for timber, ſhould never think of tranſplanting them, but ſow the acorns on the ſame ground where they are to grow; for timber of all thoſe trees which are tranſplanted is not near ſo valuable as that of the trees from acorns." (Art. QUERCUS.)

preferable

preferable to any. But in more level ſituations, we cannot allow any liberty of choice: the *drill* manner is undoubtedly the moſt eligible; and, with this method of raiſing a Wood, we begin to give our directions.

LAYING OUT LANDS FOR WOODS. But before we enter upon the immediate ſubject, it will be proper to premiſe, that, previous to the commencement of any undertaking of this nature, it would be adviſeable that the ſpot or ſpots intended to be converted into Woodland, ſhould be determined upon,—the quantity of land aſcertained,—and the whole (whether it be entire or in detached parts, and whether it be ten acres or a hundred) divided into *annual ſowings*.

THE exact number of theſe ſowings ſhould be regulated by the uſes for which the Underwood is intended. Thus, if, as in Surrey, ſtakes, edders, and hoops are ſaleable, the ſuite ought to conſiſt of eight or ten ſowings; or if, as in Kent, hop poles are in demand, fourteen or fifteen ſowings will be required; and if, as in Yorkſhire, rails be wanted, or, as in Glouceſterſhire, cordwood be moſt marketable, eighteen or twenty ſowings will be neceſſary, to produce a regular ſucceſſion of *annual falls*.

MANY

MANY advantages accrue from thus parcelling out the land into ſowings: the buſineſs, by being divided, will be rendered leſs burdenſome; a certain proportion being every year to be done, a regular ſet of hands will, in proper ſeaſon, be employed; and, by beginning upon a ſmall ſcale, the errors of the firſt year will be corrected in the practice of the ſecond, and thoſe of the ſecond in that of the third. The produce of the intervals will fall into regular courſe; and, when the whole is completed, the falls will follow each other in regular ſucceſſion.

IF it be found convenient to haſten the buſineſs, two or three diviſions may be ſown in one year, the ſeparate falls being marked by the firſt cutting. This, though by no means equal to regular ſowings, correſponding to the intended falls, is much better than hurrying over the whole buſineſs at once;—a piece of raſhneſs, which no man, who works upon an extenſive ſcale, ſhould be guilty of.

THE principal objections to raiſing Woodlands, in this progreſſive manner, is the extra trouble in fencing. However, if the ſowings lie detached from each other, the objection falls; if, on the contrary, they lie together, or in plots, let the entire plot be incloſed at once; and, if it contain a number of ſowings, ſome ſubdiviſions will be neceſſary,

ceſſary, and the annual ſowings of theſe ſubdi-viſions may be fenced off with hurdles, or other temporary contrivance. If the adjoining land to be ſown be kept under the plow, little temporary fencing will be wanted.

IT may be further neceſſary, before we enter upon the buſineſs of ſowing, to give ſome directions as to FENCING; for, unleſs this be done effectually, that will be labour loſt.

IN raiſing a Wood, from ſeeds, it is not only neceſſary to fence againſt cattle and ſheep, but againſt hares alſo, eſpecially if they be numerous. Nothing leſs than a cloſe fence is adequate to this purpoſe. Where the ſoil will admit of it, a ditch, bank, and dwarf paling, may be raiſed, in the manner already deſcribed, under the article FENCES; except that, inſtead of a ſtake-and-edder hedge, a cloſe paling ſhould be ſet upon the bank, in the following manner.

BEFORE the bank be finiſhed, the poſts, about five feet long, ſhould be put down, their lower ends being firſt *charred* (ſuperficially burnt), to prevent their decaying. One rail is ſufficient. To this the upper ends of the pales are nailed, their lower ends having been previouſly driven into the crown of the bank. The pales ſhould be about three feet long,

long, and ought to be of Oak, or the bottom parts will ſoon decay.

THE fence is the ſtronger, and more effectual, if the ditch be made on the outer ſide of it, and the paling ſet ſo as to lean outwards; but the quick ſtands a much better chance of being reared, on the inner ſide of the paling, next to the ſeedling plants: therefore, the moſt prudent method of making a fence of this kind, is to make the ditch on the outſide, without an off-ſet, leaning the paling over it, and planting the quick at the foot of the bank, on the inner ſide: it then becomes, what it ought always to be conſidered,—a part of the *Nurſery*.

THIS, however, is an expenſive fence, and is better ſuited to a ſmall than a large ſcale; and if, inſtead of the dwarf paling, a cloſe rough ſtake-and-edder hedge be ſet upon the bank, it will (provided it be well made and carefully attended to from time to time, and the *muces*, if any be made, ſtopt with rough buſhes, and ſtakes driven through them), continue to be effectual, againſt *hares*, for a conſiderable time. Againſt *rabbits*, nothing leſs than death is effectual.

AT length we come to treat particularly of the method of raiſing a Wood, upon land ſufficiently ſound,

ſound, and ſufficiently level, to be cultivated, conveniently, with the COMMON PLOW.

THE PREPARATION OF THE GROUND. If the ſoil be of a ſtiff clayey nature, it ſhould receive a whole year's fallow, as for wheat;—if light, a crop of turneps may be taken; at all events, it muſt be made perfectly clean, before the tree ſeeds be ſown; particularly from perennial root weeds: for, when once the ſeeds are ſown, all further opportunity of performing *this* neceſſary buſineſs is, in a great meaſure, loſt. If the ſituation be moiſt, the ſoil ſhould be gathered into wide lands; not high, but ſufficiently round to prevent ſurface water from lodging upon them.

THE TIME OF SOWING is either autumn or ſpring. October and November may be called the fitteſt months for the autumnal ſowing, and March for the ſpring ſowing. A man of judgement; however, will attend to the ſeaſon, and to the ſtate of his ſoil, rather than to the Calendar.

THE METHOD OF SOWING is this.—The land being in fine order, and the ſeaſon favorable, the whole ſurface ſhould be ſown with Corn or Pulſe, adapted to the ſeaſon of ſowing: if in autumn, Wheat or Rye may be choſen; in ſpring, Beans or Oats. Whichſoever of the ſpecies of Corn is adopted,

adopted, the quantity of ſeed ſhould be leſs than uſual, in order to give a free admiſſion of air, and prevent the crop from lodging.

THE ſowing of the grain being completed, that of the tree ſeeds muſt be immediately ſet about. Theſe muſt be put in, in lines, or drills, *acroſs* the lands, and in the manner beſt adapted to their reſpective natures: Acorns and Nuts ſhould be dibbled in, while Keys and Berries ſhould be ſcattered in trenches or drills, drawn with the corner of a hoe, in the manner in which garden peas are uſually ſown.

THE diſtance which we recommend to be obſerved, between the rows, is a quarter of a ſtatute rod (four feet, and one and a half inch). This may, in theory, ſeem to be an unneceſſary preciſion; but, in practice, there are many conveniencies accrue from it. In ſetting out the diſtance between the drills, a land-chain ſhould be uſed, and not a line, which is ſubject to be ſhortened or lengthened by the weather. A chain is readily divided into rods, and the quarters may be diſtinguiſhed by white paint, or other obvious marks. Stakes being driven at the ends of the drills, a line is ſtretched, to dibble or draw the trenches by *.

IF

* It may be unneceſſary to obſerve, that the drills ſhould be exactly perpendicular to the range of ſtakes, otherwiſe the meaſurement

If the plot be extenſive, *glades*, for the purpoſe of roads, ſhould be left at convenient diſtances.

The species of underwood muſt be determined by the conſumption, or demand, peculiar to the country in which it is intended to be raiſed. In Surrey, where ſtakes, edders, and hoops, are in demand, the Oak, the Hazel, and the Aſh, are eſteemed valuable, as underwoods. Upon the banks of the Wye, in Herefordſhire, Monmouthſhire, and Glouceſterſhire, where great quantities of charcoal are made for the iron forges, Beech is the prevailing underwood; but whether from choice, or from its thriving well upon thoſe bleak mountains, we cannot ſay. In Kent, where hop poles are valuable articles, the Cheſnut and the Aſh are the favorite Coppice woods. The Oak, the Aſh, the Cheſnut, the Beech, the Birch, the Wild Sorb, the Hazel, the Box, may have their peculiar excellencies, in different countries; and the choice is, of courſe, left to the perſon who has the care of the undertaking.

The species of timber has been already determined upon; the *Oak* being the only tree admiſſible

ſurement will be falſe. If the ſowings or quarters could be ſo laid out, that the drills may be of ſome determinate length, as twenty rods for inſtance, the buſineſs of meaſuring would be rendered ſtill more eaſy.

ſible in a *Wood*. The uſual ſpace allowed to timber trees, ſtanding among underwood, is thirty feet: two rods (thirty-three feet) will not be found, when the trees have fully formed their heads, too wide a ſpace. Therefore, every eighth drill, at leaſt, ſhould be ſown with acorns, dibbled in, about ſix inches aſunder *.

THE Oak and the Hazel, riſing the FIRST YEAR after ſowing, their reſpective drills will be ſufficiently diſcriminable, at harveſt; but the keys of the Aſh lie two, and ſometimes three, years in the ground, before they vegetate; and it will be convenient to have ſome diſtinguiſhing mark, in the ſtubble; in order to prevent their being diſturbed in plowing the intervals, after harveſt. To this end, if Beans be the foſtering crop, ſcatter a few Oats among the keys, the ſtubble of which will ſhew itſelf plainly, among that of the Beans; and, on the contrary, if Oats be the crop, a line of Bean ſtubble will have the ſame beneficial effect.

AT harveſt, the crop ſhould be reaped, not mown, and be carried off with all convenient care. Between harveſt and winter, a pair of furrows ſhould be laid back to back, in the middle of each in-

 terval,

* For the particulars reſpecting the propagation of the ſeveral ſpecies under conſideration, ſee their reſpective genera in the ALPHABET OF PLANTS.

terval, for the purpoſes of meliorating the ſoil for the next year's crop, and of laying the ſeedling plants dry;—while the ſtubble of the unplowed ground, on each ſide of the drills, will keep them warm during winter.

THE NEXT YEAR's crop may be Potatoes, Cabbages, Turneps; or, if the firſt was Corn, this may be Beans; or, if Beans, Wheat drilled in the Tullian manner *.

ALL that the tree drills will require, this year, will be to be kept perfectly clean, by weeding and hand hoing.

IN the ſpring of the THIRD YEAR, the drills which roſe the firſt year ſhould be looked over, and the *vacancies filled up*, from the parts where the plants are ſuperfluous: but thoſe of the Aſh ſhould be deferred until the fourth year.

THE whole ſhould afterwards be looked over, from time to time; and this, with cultivating the intervals,

* This ſpecies of culture, however, can only be practiſed in the plots and ſkreen plantations, which are mentioned in page 116: it being there determined, that lands productive of CORN and GRASS, and lying conveniently for CULTIVATION, can ſeldom be converted to WOODLAND,—merely as ſuch, and on a large ſcale,—with propriety.

intervals, and keeping the drills free from weeds, will be all that will be neceſſary, until the tops of the plants begin to interfere.

HOWEVER, if ſeedlings be wanted for the purpoſe of laying into hedges, or if tranſplanted plants be ſaleable in the country, the *ſuperfluous ſeedlings* may be drawn out of the drills, in the ſpring of the third or fourth year, and tranſplanted into ſome vacant ground.

NONE can be more proper, nor any ſo convenient, as the contiguous *intervals*, in which they may remain two or three years, without injury to the drills, and may afford a profitable crop; ſubject, however, to this diſadvantage, the ſpade muſt be made uſe of, inſtead of the plow, in cleaning the interſpaces. Nevertheleſs, a ſtock of plants of this kind are valuable, not only as articles of ſale, but for ſhelter plantations, and for filling up waſte corners of an eſtate. See p. 117.

THE FIRST CUTTING ſhould be timed by the plants themſelves. Whenever the rows of Oaks, intended for timbers, are in danger of being drawn up too ſlender for their height, by reaſon of their being too much crowded, by the interference of the rows, the whole muſt be cut down, to within a hand breadth of the ground; except the Oaks intended for ſtands, which ſhould now be ſet out, at about two rods diſtance from each other, and as nearly a quincunx, as plants moſt proper for the purpoſe will allow.

STRENGTH, cleanneſs, and upward tendency, are the criterions by which the choice of theſe ought to be determined upon. If more than one plant of this deſcription ſtand near the point deſired, it is adviſeable not to take them down, the firſt fall (provided they do not interfere too cloſely with each other), but to let them remain, in order to guard againſt accidents, and to afford a future opportunity of making a ſecond choice, when the plants are arrived at a more advanced ſtate.

THE young ſtands will require to be more or leſs *pruned:* their leaders muſt be particularly attended to, the lower ſide ſhoots taken off, and their heads reduced, in ſuch a manner, as to prevent their being rendered top-heavy.

HOWEVER, if the firſt fall of underwood be made in due time, their heads, in general, will want but little pruning; for it is not in this caſe, as in that of tranſplanting, where the roots have freſh ſhoots to make, and a freſh ſource of food to ſeek: here, they are fully prepared to ſend up the neceſſary ſupplies, and the more top there is to promote the aſcent,

aſcent, the quicker progreſs the plants will be enabled to make.

IT is, therefore, very imprudent to defer the firſt fall, until the plants be drawn up, too ſlender, to bear a well ſized top: We have known young Oaklings, raiſed in a manner ſimilar to that which is here deſcribed, drawn up ſo tall and ſlender, by injudicious treatment, as not to be able to bear the ſmalleſt top, without ſtooping under the weight of their own leaves; a ſhower of ſnow, falling without wind, bows them to the ground.

THE SECOND FALL ſhould be timed according to the *ware* which the country calls for; with this proviſo, however, that the timber ſtands be not injured, by being crouded among the under-wood; for, rather than this ſhould be the caſe, the ſecond fall ſhould take place, although the Coppice wood may not have reached the moſt profitable ſtate.

AFTER the ſecond and every SUCCEEDING FALL of underwood, the timbers ſhould be gone over, their leaders kept ſingle, and their heads ſet up, until the ſtems have reached the height of fifteen or twenty feet (more or leſs, as accidents, or their reſpective tendencies, may happen to determine), when their heads ſhould be permitted to ſpread, and take their own natural form.

So ſoon as the branches are firmly eſtabliſhed (which may happen in ten, fifteen, or twenty years from the laſt pruning, ſooner or later, according to the ſoil, ſituation, and other circumſtances), THE HEADS SHOULD BE PRUNED.

IN doing this, the leader is to be ſhortened, to check the upward growth of the tree, and the main ſtrength of the head to be thrown, as much as may be, into one principal arm; in order to obtain, with greater certainty, the important end to which *Wood* timber is more peculiarly applicable: we mean CROOKED SHIP TIMBER.

NEXT, as to RAISING A WOOD AGAINST A HANG, too ſteep to be cultivated conveniently with the *common plow*, after the Wood ſeeds are ſown; but which may, nevertheleſs, be fallowed, and brought into proper tilth by the *turn-wreſt plow*; namely, a plow which turns the furrows all one way, and which is in common uſe upon the hills of Kent and Surrey.

UNDER theſe circumſtances, the planter has it in choice, whether he will ſow ſeeds,—or put in ſeedlings,—or tranſplanted plants. If he adopt the

firſt,

first, the expence of cleaning, by hand, will fall heavy; and if the last, the labour of the Nursery will not be less burdensome. The middle path is therefore most adviseable.

THE seedling plants may, in general, be permitted to remain, until the third year, in the seed bed; by which time they will have acquired sufficient strength and stature, to struggle with the lower order of weeds, while those of a more aspiring nature may be kept under, at a reasonable expence.

THE arrangement of these plants may either be irregular, or in drills, similar to those mentioned aforegoing. After the plants are in, acorns may be dibbled in the interspaces, that success may be rendered the more secure.

THE choice of underwood, and the after management, under these circumstances, must be subject to the same rules, as under those already mentioned.

WITH respect to HANGS SO VERY STEEP, or SO STONEY, that even the turn-wrest plow cannot be used in preparing the soil, seedling plants and acorns, or other tree seeds, may be put in, without any previous preparations; except that of clearing

away

away bushes, and burning off the weeds and rough grass, with which the surface may be encumbered. In this case, the number of plants, and the quantity of acorns, should be greater, than when the ground has been prepared by a fallow.

SINCE the foregoing Remarks were written (in 1783 and 4), some favorable opportunities of collecting FARTHER INFORMATION, respecting this very important branch of Rural Economy, have occurred to us.

IN the SOUTHERN COUNTIES, we have seen the Oak rise fortuitously, or with but little assistance of the Woodman, to Timber of the first quality *. In the MIDLAND COUNTIES, we have examined Oak Woods, of different ages, which have been propagated by art, in the most simple manner: namely, that of sowing acorns with arable crops, or of setting them in the turf of grassland, and leaving the young plants to nature; and this with good success †. In the HIGHLANDS of SCOTLAND, we have observed districts of mountain surfaces covered

* Some Account of the Woodlands here alluded to, may soon appear in a Register of the RURAL ECONOMY of the SOUTHERN COUNTIES.

† See the RURAL ECONOMY of the MIDLAND COUNTIES, Vol. ii. p. 297.

covered with tree plants, of various ages and ſpecies; and this, too, with a ſucceſs, which, ſeeing the inaccurate manner in which they are frequently put in, and the neglect they afterward experience, is almoſt incredible *.

NEVERTHELESS, we ſtill remain advocates for the practice of TREATING YOUNG WOODS AS NURSERY GROUNDS. Our motives are many: by keeping the ſoil in a ſtate of tilth, and free from weeds, much time is gained in their early growth, and a ſtrong vigorous habit given to the youthful plants: by this treatment, alſo, a favorable opportunity is obtained, for removing ſupernumerary plants, for ſale, or for plots of planting, or for filling up vacancies, in parts too thinly ſtocked.

WE likewiſe retain full conviction of the propriety of TRAINING THE YOUNG TIMBER TREES OF WOODS, in ſuch manner as to render them, *with certainty*, applicable to the eſpecial purpoſes for which they are raiſed, rather than to leave them to *fortuitous circumſtances*; and ſuffer them, by ſpreading too low, to deſtroy the underwood which ſurrounds them, or, by ſhooting up too ſtraight, to fruſtrate

* See a Sketch of the RURAL ECONOMY of the CENTRAL HIGHLANDS, preſented, as a REPORT of that Diſtrict, to the BOARD OF AGRICULTURE, in Feb. 1794.

frustrate the main intention of wood timber. If straight timber be required, CLOSE GROVES, and not OPEN WOODS, are the fit places to raise it in. Land, such at least as will grow ship timber with advantage, is become too valuable to be given up, in any case, to accident or neglect.

IN Forests and other *Wastes*, whether public or appropriated, especially where the soil is of a deep clayey nature, Oaks will rise, spontaneously, from seeds that *happen* to be dropped, and whose seedling plants *happen* to be defended, by underwood or rough bushes, from the bite of pasturing animals; and some few of the plants, thus fortuitously raised, may *chance* to take the form desired by the ship carpenter: but this is all mere matter of *accident*. Even in kept woods, there may not, under the much praised system of neglect, be a sufficient crook, or a knee, fit for a first rate ship, in an acre of Woodland.

WE have repeatedly spoken our sentiments on the subject of PRUNING TIMBER TREES. To hack off a *large bough* from an *aged tree*, is a crime of the deepest dye, in the management of timber. But what relation has this mad act to the salutary operation of removing a twig from the stem of a young growing tree, or of pruning the boughs, or even of removing the leader (far above the stem), of of a tree in a youthful growing ſtate? The operations are as diſtinct as darkneſs and light, or as evil and good. In that caſe, the ſize of the wound, and the exhauſted ſtate of the tree, unite to prevent the healing; and a defect in the timber conſequently takes place: while, in this, the wound is inconſiderable, and the vigorous ſtate of the tree enables it to cicatrize the ſore, in a few months perhaps, after the operation is performed.

BY freeing the ſtems of young trees from ſide ſhoots, and by keeping their leaders ſingle, a LENGTH OF STEM is, *with certainty*, obtained; and, by afterwards checking their upright growth, and throwing the main ſtrength of the head into one principal bough (by checking, not removing, the reſt), a CROOKEDNESS of Timber is had, with the ſame certainty: and, what is equally neceſſary in SHIP TIMBER, a CLEANNESS and EVENNESS of CONTEXTURE are, at the ſame time, produced. The dangerous, and too often, we fear, fatal defect, cauſed by the decayed ſtumps of dead ſtem boughs being overgrown and hid under a ſhell of ſound timber,—a defect which every fortuitous tree-is *liable* to,—is, by this provident treatment, avoided: the timber, from the pith to the ſap, becoming uniformly ſound, and of equal ſtrength and durability.

NOTHING

NOTHING but prejudice, of the moſt inveterate kind, can reject a practice, which is founded on the moſt obvious principles of nature and reaſon; and which, in the numerous inſtances we have ſeen in hedge timber, and more particularly in the ancient avenues, which remain in every quarter of the kingdom, and which, beyond all doubt, were trained up in the manner here recommended (for without it their uniform length of ſtem could not have been had), are ſufficient proofs of its eligibility *.

UNDER a full conviction of the propriety of training up young trees, in the way beſt adapted to the purpoſes for which they are ſeverally intended, whether it be that of a wall tree, or an eſpalier, an orchard tree for fruit, or a wood tree for ſhip timber, we do not heſitate to recommend it,

* The miſchiefs done to Hedgerow and Avenue Trees, by injudicious *lopping*,—a diſgraceful treatment of Timber Trees everywhere obſervable,—have ariſen from the practice we are condemning; namely, that of taking large boughs from the ſtems of aged Trees,—theſe miſchiefs having been committed *after the trees were grown up*;—and not from the practice we are ſtrenuouſly recommending; namely, that of training young trees, *during the early ſtages of their growth*.

For other remarks on the PRUNING of TIMBER TREES,—ſee the Article HEDGEROWS, in page 102 of this Volume.

Alſo the RURAL ECONOMY of the MIDLAND COUNTIES, Vol. ii. p. 337.

it, in the ſtrongeſt terms, to every owner and manager of trees.

IN our judgement, the Royal Foreſts may not claim the merit of rational management, until men, expert in the training of timber trees for the purpoſe of building ſhips of war, be conſtantly employed in this important part of the management of National Timber *.

EVEN the LARCH, it is more than probable, may be TRAINED, with great advantage, as SHIP TIMBER ; for which it is well underſtood to be ſuperiorly adapted. In Italy, we believe, it has been applied to that purpoſe, for ages paſt. In the grounds of DUNKELD, a ſeat of the DUKE OF ATHOL, in Perthſhire, there are Larches, of conſiderable ſize, in a good form for Ship Building. Many have a CROOKEDNESS OF STEM, adapted to ribs ; and one, in particular, we obſerved with a FORKED TOP, admirably ſuited to knees. The former appeared to have ariſen from the ſtems having, while young, been in a ſtooping poſture ; and the other, from the tree having loſt its head, and two oppoſite ſide boughs having taken the office

* We are happy to find, ſince writing the above, that the SOCIETY of ARTS, in London, have, at length, taken up this ſubject. October 1795.

office of leaders *: fortuitous incidents, which art could readily copy; and, we believe, with high advantage to this iſland. For, ſhould the preſent price of bark continue, a ſupply of OAK TIMBER, for the purpoſe of building large Ships, will, it is to be feared, be greatly leſſened, if not, in ſome meaſure, cut off: a circumſtance, however, which will be the leſs regretted, by the *agricultural* intereſt, as the LARCH will flouriſh abundantly, on lands that are in a manner uſeleſs to agriculture; while the OAK, to bring it to a ſtature ſufficient for the purpoſe of conſtructing ſhips of magnitude, requires a ſoil and ſituation which may generally be applied to the uſes of huſbandry.

HAPPY, therefore, is it for this iſland, to poſſeſs two trees, oppoſite in their natures, yet equally perhaps capable of affording protection to its political independence: and, towards ſecuring ſo valuable a bleſſing, both of them ought to be reared and TRAINED with unremitting ſolicitude.

SEE more of the LARCH, in the next SECTION.

* It is obſervable of this Tree, that it bears cropping, even by cattle, with ſingular patience; ſeldom failing to renew its upward courſe, by one or more freſh leaders.

SECTION THE SECOND.

GROVES.

THE TIMBER GROVE is the prevailing *plantation* of modern time. WOODS OR COPPICES are ſeldom attempted; indeed, until of late years, clumps of Scotch Firs ſeem to have engaged, in a great meaſure, the attention of the planter.

THE SCOTCH FIR, however, is one of the laſt trees that ought to engage the attention of the Britiſh planter; and ſhould be invariably excluded from every ſoil and ſituation, in which any other timber tree can be made to flouriſh. The North aſpect of bleak and barren heights is the only ſituation in which it ought to be tolerated; and even there, the Larch is found to outbrave it. In better ſoils, and milder ſituation, the wood of the Scotch Fir is worth little, and its growth ſo licentious, as to over-run every thing which grows in its immediate neighbourhood: this renders it wholly unfit to be aſſociated with other timber trees: we, therefore,

fore, now discard it entirely from USEFUL PLANTATIONS*.

THE SPECIES OF TIMBER TREES, which we beg leave to recommend to the planter's notice, have been already mentioned, at the opening of this Chapter: They consist of

THE OAK,
THE ASH,
THE ELM,
THE BEECH,
THE LARCH, and
THE AQUATICS.

OF the tribe last mentioned, we chiefly recommend

THE POPLAR,
THE WILLOW,
THE ALDER,
THE OZIER.

To this list may be added,

THE CHESNUT,
THE WALNUT,
THE CHERRY,

as

* Nevertheless, to give variety in ornamental scenery, and as a nurse plant (if kept under due restraint), the Scotch Fir may be retained.

as ſubſtitutes for the Oak and the Beech; and the two latter, as humble repreſentatives of the princely Mahogany.

RESPECTING the *Elm*, an error prevails: MILLER and HANBURY tell us (ſpeaking more particularly of the fine-leaved ſort), that it will not flouriſh in cloſe plantations. Experience, however, leads us to be of a contrary opinion. How often do we ſee two Elms, ſtanding ſo cloſe together, that a bird could not fly through between them, yet both of them equally well ſtemmed: indeed, the ſhoots of the Elm will interweave with each other, in a manner we ſeldom ſee in any other ſpecies of tree. In groups and cloſe groves, too, we have ſeen them thrive abundantly. It is obſervable, however, that in theſe ſituations, their ſtems running up clean, and in a great meaſure free from ſide ſhoots, the timber takes a different nature, from that which is raiſed in more expoſed places;—where the lateral ſhoots being numerous, and being lopped off, from time to time, the ſtems become knotty; by which means the natural tenacity, in which the peculiar excellency of the timber of the Elm conſiſts, is conſiderably increaſed.

IN a Grove, the *Aſh* may be termed an *outſide* tree; plow beams, ſhafts, fellies, and harrow bulls, requiring a curvature, which generally takes place in the outer rows of a close plantation. The Ash, however, must not be excluded a central situation, as a straightness of grain is frequently desirable.

THE *Oak* and the *Larch* (except for the purpose of SHIP TIMBER, &c.) the *Beech* and the *Chesnut*, are *inside* trees; the carpenter, the cooper, and the turner, requiring a cleanness of grain.

WITH respect to SOIL AND SITUATION, the Elm, the Chesnut, the Walnut, and the Cherry, require a good soil and mild situation; the Aquatics should be confined to moist low grounds; and the Beech and the Larch to bleak or barren places; whilst the Oak and the Ash can accommodate themselves to almost any soil or situation; though they seldom rise to profit, on bleak and barren sites.

WE now come to the METHOD OF RAISING the several species of Grove timbers. The Oak, the Ash, the coarse-leaved Elm, the Beech, the Chesnut, the Walnut, and the Cherry, may be raised in drills, in the manner described in the preceding section, without any variation, except in the method of training. The Pines being of a hazardous nature, when in their infant state, it is adviseable to raise them in seed beds, and plant them out as seedling plants. The fine-leaved Elm must be be raiſed from layers; and the Aquatics from cuttings *.

THE METHOD OF TRAINING Grove timbers, raiſed in drills, is this: If ſeedling plants be wanted, the rows may be thinned, the third and fourth years, until the remaining plants ſtand from twelve to eighteen inches apart. This done, nothing more will be requiſite, until ſuch time as ſome kind of *ware* can be cut out; as edders, hoops, ſtakes, &c.

THE plants having reached this ſtage of their growth, the rows ſhould be gone over, every winter, and all the underling plants be cut out, within the ground (if practicable), which will, in general, kill the roots and ſave the expence of grubbing. If the remaining plants are not already too much crouded, thoſe which yet ſtruggle for the light ought to be left, to aſſiſt in drawing up, with greater certainty, thoſe which have gained the aſcendancy.

THIS conduct ſhould be obſerved, from the time of the firſt cutting, until the trees are ſet out, at diſtances beſt ſuited to their reſpective natures, and according to the accidental tendency, which

* For the method of *planting* a Timber Grove, ſee the Diviſion MANUAL LABOUR, page 33.

they happened to take, in riſing. For, in *thinning* a timber grove, little or no regard muſt be had to a regularity of diſtance at the root; an equal diſtribution of head room meriting a more particular attention.

THE ſelection ought to be directed by the ſtrength of the plants, and the uniformity of the CANOPY, taken jointly: for a chaſm in what may be called the foliage of a grove, is ſimilar to a vacancy in a coppice, or an unproductive plot in a field of corn. The leaves are as labourers; and every leaf deficient is a labourer loſt. The woodman's eye ought, therefore, to be directed towards the tops, rather than to the roots, of his trees.

THERE are other things obſervable in *thinning* a grove. If it be thinned too faſt, its upward growth will be checked, and the length of ſtem curtailed; and if, on the other hand, the thinning be neglected, or be performed too leiſurely, the plants, eſpecially in their taller ſtate, will be rendered too ſlender, and thereby become liable to laſh each other's tops, with every blaſt of wind. This evil is called *whipping of tops*, and many fine groves have been very materially injured by it. Whenever two trees are ſeen to be engaged in this conflict, one of them ſhould be taken down without

loſs

loſs of time; otherwiſe, it will probably prove fatal to them both.

IF the thinning be conducted with judgement, little *pruning* will be neceſſary; ſome, however, will be found requiſite: ſtrong maſter plants are liable to throw out ſide branches, to the annoyance of their neighbours: thoſe ſhould be taken off, in time, and all dead branches ſhould be removed, eſpecially thoſe of the pine tribe; otherwiſe, the heart of the timber will be rendered coarſe, knotty, and of a bad quality. The leaders ſhould alſo have due attention paid to them; particularly if a group of foul-headed plants happen to fall together; for, in this caſe, if nature be not aſſiſted, a timber tree will, in the end, be wanted.

THIS method of training holds good, whether the grove be raiſed from ſeeds, immediately, or from ſeedling, or other plants; and whether theſe be arranged in drills, or in the promiſcuous manner; provided the body of the grove be formed of one entire ſpecies of timber tree; for of the method of raiſing that ſpecies of grove we have hitherto been treating.

WITH regard to MISCELLANEOUS GROVES, we have ſeen ſo many evil effects, ariſing from injudicious mixtures of timber trees, that we are inclined to condemn, *as unprofitable*, all mixtures whatever. It may be argued, however, that, by aſſociating trees of different natures, the ſoil will be made the moſt of; under an idea, that each ſpecies of plant has its own favorite food: and, indeed, it is well known that corn flouriſhes after graſs, and graſs after corn; that the Aſh will thrive after the Oak, and the Oak after the Aſh, in a more profitable manner, than any one of theſe plants would do, if propagated repeatedly upon the ſame ſpot of ſoil.

THIS leads to an improvement in the method of RAISING A GROVE OF OAKS; and the ſame method is applicable to any other ſpecies of tree. Inſtead of ſowing every drill with acorns, let every ſecond be ſown with the ſeeds of a tree of a different nature; and, under ordinary circumſtances, with thoſe of the Aſh: its ſeeds are eaſily procured, and, as underwood, no tree is applicable to ſo many uſeful purpoſes.

IN this caſe, the method of training is nearly the ſame, as that already deſcribed; except that, throughout, the Aſh muſt be made ſubſervient to the Oak: if it riſe too faſt, it muſt be cut down to the ſtub, as underwood: if aſhen ſtands be left to draw up the young Oaks, they muſt be lopt, or taken down, the moment they aſpire to a ſuperiority,

periority, or give the neighbouring plants an improper tendency.

WHEN the Oaks have acquired a ſufficient length of ſtem, and have made good their canopy, the aſſiſtance of the Aſhes will be no longer wanted; nor will they be any longer valuable as underwood; they ought therefore to be entirely removed: and, if their roots be grubbed up, the Oaks will receive at once a freſh ſupply of air and paſturage.

IN bleak ſituations, a quicker growing and better feathered plant than the Aſh, affords more valuable protection: the Scotch Fir, kept under due ſubjection, is eligible in this caſe. The Furze is ſometimes made uſe of, for this purpoſe: but the plant which we wiſh to recommend, in preference to the laſt, is the Broom; as being leſs offenſive, and at the ſame time more efficacious. Its ſeeds are readily procured; its growth is rapid; it will brave the bleakeſt aſpect; and the natural ſoftneſs of its foliage renders it inoffenſive to work among, even in its talleſt and moſt crowded ſtate.

THE DUKE of PORTLAND has found, that upon the bleak ſandy ſwells of Nottingham Foreſt, the Birch affords a friendly protection to the Oak: and,

and, when we consider the easy manner in which this plant may be raised, the quickness of its growth, the shelter it gives, and its value, in many places, as an underwood, we must allow great merit in the choice.

HIS Grace's plantations being carried on upon a scale which is truly magnificent, and it being in the conducting of great undertakings, that the human invention is raised to the highest pitch, it would be unpardonable, in a work of this nature, to omit inserting the following Letter from Mr. SPEECHLY, his Grace's Gardener, to Dr. HUNTER, Editor of a late edition of EVELYN'S *Sylva*, describing the manner in which these plantations have been conducted.

WE introduce it, in this place, as the style of planting it describes is peculiarly adapted to raising GROVES against HANGS, or acclivities of hills. The candour contained in the Letter itself precludes the necessity of apprizing our readers, that it is not calculated for a STRONG LEVEL COUNTRY, nor for raising Woods, in any soil or situation.

——'FEW Noblemen plant more than his 'Grace the Duke of Portland; and I think I may 'say, without vanity, none with greater success. 'But as no man should think of planting in the very 'ex-

'extensive manner that we do, before he is pro-
'vided with well-stocked nurseries, it may not be
'amiss, before I proceed further, to give a short
'sketch of that necessary business, as also to inform
'you of the soil and situation of our seat of plant-
'ing. The greatest part of our plantation is on
'that soil which in Nottinghamshire is generally
'distinguished by the name of Forest land. It is a
'continuation of hills and dales; in some places
'the hills are very steep and high; but in general
'the ascents are gentle and easy.

'THE soil is composed of a mixture of sand and
'gravel; the hills abound most with the latter, and
'the vallies with the former, as the smaller particles
'are by the wind and rains brought, from time to
'time, from the high grounds to the lower. It is
'on the hilly grounds we make our plantations,
'which in time will make the vallies of much
'greater value, on account of the shelter they
'will afford.

'AFTER his Grace has fixed on such a part of
'this Forest land as he intends to have planted,
'some well situated valley is chosen (as near the
'center of the intended plantations as may be) for
'the purpose of a nursery; if this valley is sur-
'rounded with hills on all sides but the south, so
'much the better. After having allotted a piece
'of

'of ground, conſiſting of as many acres as is con-
'venient for the purpoſe, it is fenced about in ſuch
'a manner as to keep out all obnoxious animals.
'At either end of the nurſery are large boarded
'gates, as alſo a walk down the middle, wide
'enough to admit carriages to go through, which
'we find exceedingly convenient when we remove
'the young trees from thence to the plantations.
'After the fence is completed, the whole is
'trenched (except the walk in the middle) about
'twenty inches deep, which work may be done
'for about three pounds ten ſhillings, or four
'pounds, per acre, according as the land is more
'or leſs gravelly; this work is beſt done in the
'ſpring, when the planting ſeaſon is over. If,
'after the trenching, two or three chaldrons of
'lime be laid on an acre, the land will produce an
'excellent crop either of cabbages or turneps,
'which being eaten off by ſheep in the autumn,
'will make the land in fine order for all ſorts of
'tree ſeeds: but as the Oak is the ſort of tree we
'cultivate in general, I ſhall confine myſelf parti-
'cularly to our preſent method of raiſing and
'managing that moſt valuable ſpecies. In the
'autumn, after the cabbage or turneps are eaten
'off, the ground will require nothing more than a
'common digging. So ſoon as the acorns fall,
'after being provided with a good quantity, we
'ſow them in the following manner: Draw drills
'with

'with a hoe in the ſame manner as is practiſed for
'peaſe, and ſow the acorns therein ſo thick as
'nearly to touch each other, and leave the ſpace of
'one foot between row and row, and between every
'fifth row leave the ſpace of two feet for the alleys.
'While the acorns are in the ground, great care
'muſt be taken to keep them from vermin, which
'would very often make great havock among the
'beds, if not timely prevented. Let this caution
'ſerve for moſt other ſorts of tree ſeeds.

'AFTER the acorns are come up, the beds will
'require only to be kept clean from weeds until
'they want thinning; and as the plants frequently
'grow more in one wet ſeaſon, where the ſoil is
'tolerably good, than in two dry ones, where the
'ſoil is but indifferent, the time for doing this is
'beſt aſcertained by obſerving when the tops of
'the rows meet. Our rule is to thin them then,
'which we do by taking away one row on each
'ſide the middlemoſt, which leaves the remaining
'three rows the ſame diſtance apart as the breadth
'of the alleys. In taking up theſe rows, we ought
'to be anxiouſly careful neither to injure the roots
'of the plants removed, nor of thoſe left on each
'ſide. The reſt of the young Oaks being now
'left in rows at two feet apart, we let them again
'ſtand until their tops meet; then take up every
'other row, and leave the reſt in rows four feet

'aſunder,

'asunder, until they arrive to the height of about
'five feet, which is full as large a size as we ever
'wish to plant. In taking up the two last sizes,
'our method is to dig a trench at the end of each
'row full two feet deep, then undermine the plants,
'and let them fall into the trench with their roots
'entire.

'AND here let me observe, that much, very
'much, of their future success, depends on this
'point of their being well taken up. I declare
'that I should form greater hopes from one hun-
'dred plants well taken up and planted, than from
'ten times that number taken up and planted in a
'random manner; besides, the loss of the plants
'makes the worst method the most expensive.

'BUT before I leave this account of our method
'of raising Oaks, I shall just beg leave to observe,
'that we are not very particular in the choice of
'acorns; in my own opinion, it matters not from
'what tree the acorns are gathered, provided they
'are good; for although there seems to be a
'variety of the English Oak, in respect to the form
'of the leaf and fruit, also their coming into leaf at
'different seasons, with some other marks of dis-
'tinction, yet I am of opinion that they will all
'make good timber trees if properly managed.
'It is natural to suppose, that a tree will grow low
'and

'and ſpreading in a hedge row; on the contrary,
'it is very improbable that many ſhould grow ſo
'in a thick wood, where, in general, they draw one
'another up ſtraight and tall. And I have ob-
'ſerved, that the ſame diſtinctions hold good
'amongſt our large timber trees in the woods, as
'in the low-ſpreading Oaks in the hedge rows.

'Though I have not, as yet, taken notice of any
'other ſort of tree but the Oak, yet we have a
'great regard for, and raiſe great quantities of,
'Beech, Larch, Spaniſh Cheſnut, Weymouth
'Pine, and all ſorts of Firs, the Scotch excepted,
'as well as many other kinds, by way of thick-
'ening the plantations while young; among which
'the Birch has hitherto been in the greateſt eſti-
'mation, it being a quick growing tree, and taking
'the lead of moſt other ſorts on our poor foreſt
'hills; and as we have an inexhauſtible ſpring of
'them in the woods, where they riſe of themſelves
'in abundance from ſeed, we at all times plant
'them plentifully of different ſizes. As to the
'Elm and Aſh, we plant but few of them on the
'Foreſt, though we raiſe great quantities of both,
'but particularly the Aſh, which being an uſeful
'wood (but a bad neighbour among the Oaks),
'we plant in places apart by itſelf. I ſhall diſmiſs
'this ſubject concerning the management of our
'nurſeries, after ſaying a word or two relating to
'pruning:

'pruning: we go over the whole of the young 'trees in the nurſery every winter; but in this we 'do little more than ſhorten the ſtrong ſide ſhoots; 'and take off one of all ſuch as have double leads.

'Having thus pointed out the mode of forming 'and managing our nurſeries, I ſhall now proceed 'to the plantations. The ſize of the plantations, 'at firſt beginning, muſt be in proportion to the 'ſtock of young trees in the nurſery; for to under-'take to plant more ground than we have young 'trees to go through with for thick plantations, 'would turn to poor account on our foreſt hills. 'We always plant thick, as well as ſow plenti-'fully at the ſame time, provided it be a ſeaſon in 'which acorns can be had; ſo that all our plan-'tations anſwer in a few years as nurſeries to ſuc-'ceeding plantations.

'As to the form of the plantations, they are 'very irregular; we ſometimes follow a chain of 'hills to a very great diſtance; ſo that what we 'plant in one ſeaſon, which perhaps is ſixty, 'eighty, and ſometimes an hundred acres, is no 'more than a part of one great deſign.

'If the ground intended to be planted has not 'already been got into order for that purpoſe, it 'ſhould be fenced about at leaſt a twelvemonth 'before

'before it is wanted to plant on, and immediately
'got into order for a crop of turnips; two chal-
'drons of lime being laid on an acre will be of
'great ſervice, as it will not only be a means of
'procuring a better crop of turnips, but will bind
'the land afterwards, and make it fall heavy,
'which is of great uſe when it comes to be planted,
'as ſome of the foreſt land is ſo exceedingly
'light as to be liable to be blown from the roots of
'the young trees after planting: therefore we find
'it to be in the beſt order for planting about two
'years after it has been plowed up from paſture,
'before the turf is too far gone to a ſtate of decay.
'It will be neceſſary to have a part of the turnips
'eaten off ſoon in the autumn, in order to get the
'ground into readineſs for early planting; for we
'find the forward planting generally ſucceeds the
'beſt.

'AFTER the turnips are eaten off, we plow the
'ground with a double-furrow trenching plow
'made for that purpoſe, which, drawn by ſix
'horſes, turns up the ground completely to the
'depth of twelve or thirteen inches: this deep
'plowing is of great ſervice to the plants at the
'firſt, and alſo ſaves a great deal of trouble in
'making the holes. After the plowing is finiſhed,
'we divide the ground into quarters for the planting
'by ridings. It will be a difficult matter to deſ-

 'cribe

'cribe the laying out the ground for this purpoſe, 'eſpecially where there is ſuch a variety of land 'as we have on the foreſt; much depends on the 'taſte of the perſon employed in this office. Be-'tween the hills, towards the outſides of the plan-'tations, we frequently leave the ridings from 'ſixty to an hundred yards in breadth, and con-'tract them towards the middle of the woods, to 'the breadth of ten or twelve yards; and on the 'tops of the hills where there are plains, we fre-'quently leave lawns of an acre or two, which 'makes a pleaſing variety.

'In ſome of them we plant the Cedar of Liba-'nus at good diſtances, ſo as to form irregular 'groves; and this ſort of tree ſeems to thrive to 'admiration on the foreſt-land. On the outſides of 'the woods, next to the ridings, we plant Ever-'greens, as Hollies, Laurels, Yews, Junipers, &c. 'and theſe we diſpoſe of in patches, ſometimes 'the ſeveral ſorts entire, at other times we inter-'mix them for variety; but not ſo as to make a 'regular ſcreen or edging. Our deſign in the diſ-'tribution of theſe plants, is to make the outſides of 'the woods appear as if ſcalloped with Evergreens 'intermixed ſometimes with rare trees, as the '*Liriondendron Tulipifera*, or Virginian Tulip-'tree, &c.

'AFTER

'AFTER the ground is laid out into quarters for 'planting, we aſſign certain parts to Beech, Larch, 'Spaniſh Cheſnuts, &c. Theſe we plant in irregu-'lar patches here and there, throughout the planta-'tions, which, when the trees are in leaf, has the moſt 'pleaſing effect, on account of the diverſity of 'ſhades; eſpecially in ſuch parts of the foreſt 'where four, five, and ſometimes more of the large 'hill-points meet in the ſame valley, and tend, as it 'were, to the ſame center.

'AFTER thoſe patches are planted, or marked 'out for that purpoſe, we then proceed to the 'planting in general. We always begin with 'planting the largeſt young trees of every ſort, and 'end our work with thoſe of the ſmalleſt ſize: 'were we to proceed otherwiſe, the making a 'hole for a larger ſized tree, after the ſmall ones 'are thick planted, would cauſe the greateſt con-'fuſion.

'BIRCH is generally the ſort of tree we make 'our beginning with, which we find will bear to be 'removed with great ſafety, at the height of ſix or 'ſeven feet, though we commonly plant rather 'under than at that ſize. This ſort of tree we 'are always ſupplied with from our plantations of 'five or ſix years growth. But before I proceed 'to the taking them up, it will be proper to in-

'form you, that in the planting ſeaſon we divide
'our hands into four claſſes, which we term
'Takers-up, Pruners, Carriers, and Planters: and
'here I ſhall deſcribe the ſeveral methods of doing
'this work.

'FIRST, in taking up we have the ſame care to
'take up with good roots in the plantations, as was
'recommended in the nurſery, though we cannot
'purſue the ſame method; but in both places, ſo
'ſoon as the plants are taken up we bed them in
'the ground in the following manner: Dig a trench
'at leaſt fifteen inches deep, and ſet the young
'trees therein with their tops aſlant, covering the
'roots well as we go along, and almoſt half way
'up the ſtem of the plants, with the earth that
'comes out of a ſecond trench, which we fill in the
'like manner, and ſo proceed on till we have a
'load more or leſs in a heap, as may be convenient
'to the place from whence they were taken. In
'our light ſoil this trouble is but little, and we
'always have our plants ſecure, both from their
'roots drying, and their ſuffering by froſt. We
'have a low-wheeled waggon to carry them from
'the heaps, where they are bedded, to the pruners,
'and generally take two loads every other day.
'When they arrive, the planters, pruners, &c. all
'aſſiſt to bed them there, in the ſame manner as
'before deſcribed. We have a portable ſhed for
'the

'the pruners to work under, which is alſo conve-
'nient for the reſt of the work-people to take
'ſhelter under in ſtormy weather. From the
'above heaps the plants are taken only ſo faſt as
'they are wanted for pruning, which work we
'thus perform: Cut off all the branches cloſe to
'the ſtem to about half the height of the plant,
'ſhortening the reſt of the top to a conical form in
'proportion to the ſize of the plant; and in prun-
'ing of the roots, we only cut off the extreme parts
'that have been bruiſed by the taking up, or ſuch
'as have been damaged by accident, wiſhing at all
'times to plant with as much root as can be had.

'As ſoon as they are pruned they are taken to
'the planters, by the carriers, who are generally a
'ſet of boys, with ſome of the worſt of the labour-
'ers. The planters go in pairs; one makes the
'holes, and the other ſets and treads the plants
'faſt, which work they commonly do by turns.
'In making of the holes we always take care to
'throw out all the bad ſoil that comes from the
'bottom: if the planting be on the ſide of a hill,
'we lay the bad ſoil on the lower ſide of the hole,
'ſo as to form a kind of baſon; for without this
'care our plants would loſe the advantage of ſuch
'rains as fall haſtily. We at all times make the
'holes ſufficiently large, which is done with great
'eaſe after our deep plowing.

'BEFORE we ſet the plant, we throw a few ſpade-
'fuls of the top ſoil into the hole, ſetting the plant
'thereon with its top rather inclining to the weſt;
'then fill up the hole with the beſt top ſoil, taking
'care that it cloſes well with the roots, leaving no
'part hollow. When the hole is well filled up,
'one of the planters treads and faſtens the tree
'firmly with his feet, while his partner proceeds to
'make the next hole.

'THE faſtening a tree well is a material article
'in planting; for if it once becomes looſe, the
'continual motion which the wind occaſions, is ſure
'to deſtroy the fibres as faſt as they are produced,
'which muſt end in the deſtruction of the plant, if
'not prevented. It is to guard againſt this inconve-
'niency that we take off ſo much of the top, as
'has been deſcribed in the article of pruning.

'WE plant about three or four hundred Birches
'of the large ſize on an acre, and nearly the ſame
'number of the firſt-ſized Oaks; we alſo plant
'here and there a Beech, Larch, Spaniſh Cheſnut,
'&c. excluſive of the patches of the ſaid ſorts of
'trees before planted. We then proceed to plant
'plentifully of the ſecond and leſſer-ſized Oaks;
'and laſt of all a great number of the ſmall
'Birches, which are procured from the woods at
'about three ſhillings or three ſhillings and ſixpence
'per

'per thousand: these we remove to the succeed-
'ing plantations after the term of five or six years.
'Of the several sizes of the different kinds of
'trees, we generally plant upwards of two thou-
'sand plants upon an acre of land, all in an irregu-
'lar manner.

'AFTER the planting is finished, we then sow
'the acorns (provided it be a season that they can
'be had) all over the plantation, except amongst
'the Beech, Larch, &c. in the aforesaid patches.
'Great care should be taken to preserve the acorns
'intended for this purpose, as they are very sub-
'ject to sprout, especially soon after gathering;
'the best method is to lay them thin in a dry airy
'place, and give them frequent turnings. We
'sow these acorns in short drills of about a foot
'in length, which work is done very readily by
'two men, one with the acorns, the other with a
'hoe for the purpose of making the drills and
'covering the seed.

'WE are of opinion that the plants produced
'from these acorns will at last make the best trees;
'however, I will not pretend to say how that may
'be, as the Oaks, transplanted small, grow
'equally well for a number of years: but it is
'probable that a tree with its tap-root undisturbed
'may, in the end, grow to a much larger size.

'AFTER the whole is finiſhed to a convenient
'diſtance round the pruners, we then remove their
'ſhed to a ſecond ſtation, and there proceed in the
'like manner; and ſo on till the whole be finiſhed.

'IT would be well to get the planting done by
'the end of February, eſpecially for trees of the
'deciduous kind; but from the diſappointments
'we meet with, occaſioned by the weather, we are
'ſometimes detained to a later ſeaſon.

'I HAVE ſeveral times made trial of twelve or
'fourteen kinds of American Oaks ſent over to
'his Grace in great quantities. I ſowed them in
'the nurſery, and alſo in the beſt and moſt ſhel-
'tered parts of the plantations. In both places
'they come up very plentifully; but I now find
'that ſeveral of the ſorts will not ſtand the ſeverity
'of our winters, and thoſe that do make ſo ſmall
'a progreſs as to promiſe no other encouragement
'than to be kept as curioſities.

'TOWARDS the end of April, when the ground
'is moiſt, it will be a great ſervice to go over the
'whole plantations, and faſten all ſuch trees as are
'become looſe ſince their planting: after this,
'nothing more will be required till the month of
'June, when we again go over the whole with
'hoes, cutting off only the tall growing weeds;
'for

‘ for the ſooner the ground gets covered with graſs,
‘ in our light ſoil, ſo much the better.

‘ I OWN there is ſomething ſlovenly in the ap-
‘ pearance of this method, and on ſome lands I
‘ would recommend keeping the ground clean
‘ hoed for ſome time at firſt, as alſo planting in
‘ rows, which in that caſe would be neceſſary.
‘ More than once I have tried this method on our
‘ foreſt hills, and always found, after every hoeing,
‘ that the ſoil was taken away by the ſucceeding
‘ winds into the valleys.

‘ BESIDES this inconvenience, the reflection of
‘ our ſandy ſoil is ſo very great, that we find the
‘ plants ſtand a dry ſeaſon much better in our pre-
‘ ſent method, than in the former: and whoever
‘ fancies that graſs will choak and deſtroy ſeedling
‘ Oaks, will, after a few years trial, find himſelf
‘ agreeably miſtaken: I have even recommended
‘ the ſowing the poorer parts of the hills with furze
‘ or whin ſeed, as ſoon as they are planted: we
‘ have ſometimes permitted the furze to grow in
‘ the plantations by way of ſhelter for the game,
‘ which though it ſeems to choak and overgrow
‘ the Oaks for ſome time, yet after a few years
‘ we commonly find the beſt plants in the ſtrongeſt
‘ beds of whins. This ſhews how acceptable
‘ ſhelter is to the Oak whilſt young; and expe-
‘ rience

'rience ſhews us, that the Oak would make but a
'ſlow progreſs on the foreſt hills for a number of
'years at the firſt, were it not for ſome kind nurſes;
'and the Birch ſeems to anſwer that purpoſe the
'beſt, as I have already obſerved.

'THE ſeveral ſorts of Fir trees, from appearance,
'ſeem to promiſe a greater ſhelter; but on the
'foreſt land they do not grow ſo faſt as the former,
'and what is worſe, the Oak will not thrive under
'them, as they do immediately under the Birch.

'WHERE a plantation is on a plain, a ſcreen of
'Firs for its boundary is of ſingular uſe, but the
'ſituation of the foreſt land denies us this ad-
'vantage.

'WE continue to cut down the tall growing
'weeds two or three times the firſt ſummer, and
'perhaps once the next, or ſecond ſeaſon after
'planting; which is all that we do in reſpect to
'cleaning. The next winter after planting, we
'fill up the places with freſh plants where they
'have miſcarried; after which there is little to be
'done till about the fourth or fifth year; by which
'time the ſmall-ſized Birch, and ſeedling Oaks,
'will be grown to a proper ſize for tranſplanting:
'in the thinning of theſe due care muſt be had
'not to take too many away in one ſeaſon, but,
'being

'being properly managed, there will be a supply
'of plants for at least half a dozen years to come.

'ABOUT the same time that the lesser-sized
'Birch wants thinning, the large ones will require
'to have their lower branches taken off, so as to
'keep them from injuring the Oaks; and this is
'the first profit of our plantations, the Birch wood
'being readily bought up by the broom makers.
'This pruning we continue as often as required,
'till the Birches are grown to a sufficient size to
'make rails for fencing; we then cut them down
'to make room for their betters.

'BY this time the Oaks will be grown to the
'height of twelve or fourteen feet, when they draw
'themselves up exceedingly fast: each plant seems
'as it were in a state of strife with its neighbour,
'and in a strict sense they are so, and on no other
'terms than life for life; and he whose fate it is to
'be once over-topped, is soon after compelled to
'give up the contest for ever.

'AFTER the Birches are cut down, there is
'nothing more to be done but thinning the Oaks,
'from time to time, as may be required, and cut-
'ting off their dead branches as frequently as may
'be necessary. We are very cautious in doing
'the former, knowing well that if we can but once

'obtain

'obtain length of timber, time will bring it into
'thickneſs; therefore we let them grow very cloſe
'together for the firſt fifty years.

'AND here it may not be improper to obſerve
'the progreſs the Oak makes with us, by deſcribing
'them in two of our plantations, one of twenty-
'eight, the other of fifty years growth. In the
'former they are in general about twenty-five or
'twenty-ſix feet in height, and in girth about
'eighteen inches: the trees in the latter, planted
'in 1725, are ſomething more than ſixty feet in
'height, and in girth a little above three feet; and
'theſe trees are in general about fifty feet in the
'bole, from which you will eaſily conceive the
'ſmallneſs of their tops, even at this age.

'IT would be a difficult matter to deſcribe their
'farther progreſs with any degree of certainty,
'therefore let it ſuffice to make this laſt obſer-
'vation on them in their mature ſtate.'

'*Welbeck*, 16 *June*, 1775.

THIS valuable Paper does Mr. SPEECHLEY great credit. On the ſpecies of *Planting*, which he here deſcribes, it is in itſelf a Treatiſe.

BUT it ſtrikes us forcibly, that much of the expence of the great and laudable undertaking, which is is the ſubject of it, might have been ſaved, by *ſowing* the tree ſeeds on the ſites to be wooded.

WE are fully aware of the impropriety of keeping, in a looſe pulverous ſtate, the intervals of tree plants, on a blowing ſand, and in an expoſed ſituation; but, in the method we have mentioned, as being practiſed in the Midland Counties, of ſowing the tree ſeeds with corn, or of depoſiting them in the turf of graſs land, this ill effect of light ſandy land is avoided.

WERE we to recommend a practice for the Sand hills of Sherwood Foreſt, or for any other ſite of a ſimilar nature, it would be that of preparing the ſoil, by a clean fallow, for rye; ſowing or dibbling in the tree ſeeds; mixing thoſe of the timber trees and the nurſe plants promiſcuouſly, or in alternate drills; and, having previouſly guarded the ſeedling plants, by ſufficient fences, to let them remain, under the ſhelter of the ſtubble and the weeds that might ſpring up, until the plants were ſufficiently conſpicuous, to aſcertain their ſuperabundance or deficiency: and, having then filled up the vacancies, with the ſupernumerary plants of ſuch parts as might be too thickly ſtocked,—ſetting out the whole at proper diſtances, as a field of turnips or of ſeed rape is ſet out,—let them remain, until future thinnings or cutting be required.

ON

On all ſoils, and perhaps in every caſe where the ſurface is occupied by a free clean ſward, depoſiting the tree ſeeds, particularly acorns, among the roots of the graſs, will, we are of opinion, be found the moſt eligible practice. This may be done, either by raiſing up a tongue of the ſward, and putting the acorn under it, as was practiſed with ſucceſs, in one inſtance, in Warwickſhire *; or by inſerting them with the common dibble; or by preſſing them into the turf, while wet, with a roller, or with the foot. The acorn will riſe, the firſt ſummer, eight or ten inches high, and ſtrike down a root a foot or more in depth; thus bidding defiance to the graſſes and moſt of the herbaceous tribes †. However, in bleak ſituations, where fibrous-rooted nurſe plants may be required, this mode of cultivation may be the leſs eligible.

There is one circumſtance obſervable, in SEMINATING THE OAK, which is not, we believe, ſufficiently attended to. It ſhould never be attempted (unleſs in extraordinary caſes) when acorns are not abundant. It is not the extra coſt of acorns, or the difficulty of procuring them, ſo much as the difficulty of preſerving them from vermin, which

* See the RURAL ECONOMY OF THE MIDLAND COUNTIES, Vol. ii. p. 298.

† See as above, page 308.

which renders this precaution requiſite. In a plentiful year, when every wood and every Hedge-row is ſtrewed with acorns, thoſe which are lodged in the ſoil are leſs liable to their ravages.

IT may be needleſs to obſerve, that the greater quantity there is ſown, in any one place, the leſs will be the proportional damage. Hence, fifty or a hundred buſhels, ſown in the field, are more likely to be preſerved, than a few in a nurſery bed. And, for a ſimilar reaſon, it may be prudent to ſow the margins of a field thicker than the area, where fewer enemies may be expected.

IT now only remains to mention the PLANTATIONS OF THE HIGHLANDS of SCOTLAND,—which have, of late years, ſpread with aſtoniſhing rapidity. There are few men of large property, within the Highlands, or on their margins, who have not ſet out their millions of tree plants, and converted, perhaps, their hundreds of acres to a ſtate of woodland; and this, in places where, twenty years ago, not a ſtick was ſeen ſtanding.

ABOUT fifty years ſince, much planting was done on two of the principal eſtates of the Highlands, thoſe of ATHOL and BREADALBANE. But the ſpirit did not diffuſe itſelf, until many years after that time.

THE

THE *species of plantation*, found in this quarter of the iſland, is uniformly the GROVE, on the rugged ſides, and on the lower ſtages, of the mountains. The ſite is generally too ſteep, and always too rough and ſtoney, to admit of being prepared with the plow. And the ſurface being generally covered with heath, or other coarſe mountain plant, *ſowing* the tree ſeeds on the ſites, is ſeldom, we believe, attempted: *planting* being the univerſal practice; at leaſt, ſo far as has fallen within our own notice or information.

THE *ſpecies of plant* has been, too generally, the NATIVE FIR; except on the lower, better-ſoiled ſites, where the OAK, and other DECIDUOUS TREES have been propagated. Of late years, however, the LARCH has been the favorite plant; it having been found to thrive on the moſt barren ſoils, and in the bleakeſt and moſt expoſed ſituations, in a manner ſuperior even to the native Pine! And its timber has been proved to be of infinitely greater value. In water work, as well as in ground work,—the beſt teſts of the quality of timber,—the Larch has been found ſingularly durable.

THE *method of planting* varies, with the age and the nature of the plant, with the ſtate of the ground, and with the ſkill of the planter.

SEED-

SEEDLING PLANTS are put in, with a dibble, or with a chop or chops of a ſpade, in the freeſt and beſt parts of the ſoil. But, for nurſery plants, which, when the ſurface is much encumbered with tall heath, are often planted, we underſtand *,— holes are made with the ſpade; firſt ſtriking off, beneath the ſurface, the heath and other natural produce; and, then, digging a pit, proportioned to the ſize of the given plant.

ON planting the *common Fir*, in theſe holes, the mold that has been raiſed is reduced with the ſpade, and returned into the pit; acroſs the center of which a deep wide gaſh or cleft is opened, with the ſpade ſtruck down to the bottom of the hole; and the roots of the plant thruſt into this cleft; which is cloſed by treading the ſoil on either ſide of it; the whole operation being, in this caſe, performed by the ſame perſon.

BUT, in planting the *Larch* and other trees, in theſe pits, two perſons are employed; the one to hold the plant, the other to reduce the mold and bed the roots, in the ordinary manner.

THIS extra coſt of planting may have determined ſome in favor of the Fir; but, when the

* Not having remained in the Highlands, during the planting ſeaſon, we ſpeak here, from information.

ſuperior value of the Larch is taken into the account, the ſaving will become, in the end, a ſerious loſs.

A STRIKING proof of the SUPERIORITY of the LARCH, in *waterworks*, occurred on the eſtate of Athol. A weir, or river dam, which, while conſtructed with Oak, required to be renewed or repaired, every four or five years, was formed with Larch; and, in 1792, had ſtood nine or ten years; the timber, then, remaining in a ſound firm ſtate. In the character of *gate poſts*, too, the Larch has been found to be ſingularly durable.

IT is ſomewhat aſtoniſhing, that, ſeeing the ſucceſs of the LARCH on the eſtate of Athol, during the laſt half century, its propagation ſhould not have ſpread more rapidly. There is probably more Larch timber, now, on that eſtate, than in the reſt of the iſland. In 1792, His Grace the Duke of Athol (we ſpeak from the higheſt authority) was poſſeſſed of a thouſand Larch trees, then growing on his eſtates of Dunkeld and Blair only, of not leſs than two to ſour tons of timber each; and had, at that time, a million Larches, of different ſizes, riſing rapidly on his eſtate. Theſe alone, IF PROPERLY TRAINED*, would ſupply the Britiſh navy with ſhip timber, for

* See page 143.

a length

a length of years. Should the ſpirit of propagating the Larch continue, nay, were it to expire at this time (1795), the Highlands of Scotland, alone, will henceforward be able to furniſh the whole commerce of the Iſland with timber for its ſhipping.

IT is not, therefore, on the mountains of Scotland, we *now* recommend, with eager ſolicitude, the propagation of the Larch. We have lands in England, and nearer to our ſhip yards, which will probably pay a hundred fold in Larch, compared with any other crop they are capable of producing. We mean, generally, the barren heathy ſurfaces which occupy no inconſiderable portion of the kingdom; but more particularly, the ſingularly INFERTILE FLATS OF HEATH, in the SOUTHERN COUNTIES of Surrey, Suſſex, and Hampſhire: lands which, at preſent, lie in a manner uſeleſs to the community; yet on which we have ſeen the Larch riſing with luxuriance!

IF theſe waſtes ſhould be planted progreſſively, with the Larch, and their produce properly TRAINED FOR SHIP BUILDING, the ſeveral yards of Portſmouth, Deptford, Chatham, &c. to which it might, at all times, be ſafely and readily conveyed by inland navigation, could not experience a want of timber, for ages to come.

WERE other wastes of a similar nature, lying in different parts of the Island, particularly the MOUNTAIN BROWS of the ENGLISH HIGHLANDS, in Yorkshire, Westmoreland, &c. and also the Cornish and Devonshire Mountains, with the Welch and Salopian Hills, together with other barren heights, at present merely blank surfaces, which lie a disgrace to the POLITICAL, as well as the RURAL ECONOMY of the kingdom, converted to the same valuable purpose, a supply of foreign timber might, in half a century, become unnecessary; and this, without any, or but an inconsiderable, abridgment of Agricultural produce.

THE Larch not only flourishes in bleak and barren sites, but ENCREASES with a RAPIDITY unknown to every other durable wood. In the grounds of BLAIR OF ATHOL, we measured a Larch, which, at five feet high, girted upwards of eight feet, and contained by estimation four tons of timber; which Larch, by the indisputable evidence of a person who remembered its being planted, was not, at the time we measured it, 1792, fifty-four years old. And, at DUNKELD, we measured another, of very little more than fifty years old, which girted, at the same height, eight feet six inches; its height near a hundred feet, and its contents from four to five tons of timber.

WE

WE are not apt to be carried away by novel ideas, and upſtart practices; on the contrary, ſeeing the falſe baſis on which they too frequently reſt, we may ſometimes, perhaps, remain in doubt, when we ought to decide: but, believing this to be the ſafer conduct, we adhere, and mean to adhere, to our principle. Nevertheleſs, in the multitude of evidences which have occurred to us, in favor of the tree now under notice, we find ſufficient ground for deciſion; and we think it right to loſe no time, in recommending it to the attention of men of property, in every diſtrict of the Iſland, in which barren heathy lands are found.

SECTION THE THIRD.

COPPICES.

AFTER what has been recommended, in the foregoing Sections, with regard to the raiſing of WOODS and GROVES, ſcarcely anything remains to be added, here, reſpecting COPPICES; the proper culture being ſimilar, in the ſeveral caſes.

THE MODERN COPPICES OF KENT (we ſpeak more particularly of the diſtrict of Maidſtone), raiſed for the purpoſe of hop poles, are chiefly of *Aſh* and *Cheſnut*; which are generally *cultivated in*

 rows;

rows; the intervals being kept clean, as thoſe of hop grounds; and the profits ariſing from them is almoſt incredible.

ONE particular in the Kentiſh practice deſerves notice. To keep the *intervals free from weeds*, and the *ſoil mellow*, they are thickly *covered with "hop bines*,"—the ſtalks of hops as freed from the poles,—and with good effect. When theſe bines have performed their office, and are become ſufficiently tender for the operation, they are dug under as manure. Furze, Broom, or rough buſhes, might be uſed in the ſame intention.

IN RAISING A COPPICE, as in cultivating any other ſpecies of woodland, the firſt buſineſs is to *regulate the plants*; to ſet them out, at proper diſtances, where they are too thick, and to fill up the vacant ſpaces with the ſupernumerary plants.

SOMETHING, too, may afterwards be done, by judicious *thinnings*; but leſs in coppices, than in the other two ſpecies of woodland. However, where the demand of the country calls for the larger articles of coppice ware, many ſtakes, binding rods, &c. may be cut out, with advantage to the free-ſhooting plants, left to grow up, to ſupply the markets of the given diſtrict; which will ever determine the SPECIES OF COPPICE WOOD. See p. 119, and the SPECIES OF UNDERWOOD, p. 130.

ONE ſpecies of coppice wood, however, requires to be particularly noticed; as its uſes are adapted to every diſtrict: namely, the OZIER; which, in low moiſt ſituations, may be cultivated, on a ſmall ſcale at leaſt, with great advantage to every farm; for binders, thatching rods, hurdles, edders, ſtakes, rake handles, ſithe handles, and other utenſils of huſbandry, and for poles and rails of almoſt any length *.

IN cultivating the Ozier, as a coppice wood, on moiſt moory ſites, the firſt ſtep is to throw the ſoil

N 4 into

* WILLOW POLLARDS are uſeful in the ſame intention; but do not afford ſuch a length and cleanneſs of ſtem, as a cloſe coppice. Nevertheleſs, they are planted, with great profit, by the ſides of brooks and rivulets, paſſing through meadowy and marſhy grounds, in many parts of the Iſland; and might, in many others, be planted with equal benefit.

An error, too frequently committed, in planting Willow poles for Pollards, is to ſet them within the banks of the rivulet or brook; to the future injury of its channel: a practice which no commiſſion of ſewers, or manor inqueſt, ſhould ſuffer.

The proper ſituation for theſe Pollards is ſome feet, not leſs than half a rod, from the brink of the channel: a ſituation, which the *Salix* tribe in general prefer; the roots ſoon reach the moiſture, and thus gain a double range of paſturage. In this ſituation, too, the trees afford a ſalutary ſhade to catlle in hot weather, without danger to themſelves, or injury to the channel, or its banks.

into beds, ſo as to lay the ſurface ſufficiently dry; the Ozier diſliking an unſound ſituation.

This work ſhould be done in autumn, when the ſoil, having had all the ſummer to grow firm in, will ſtand to the ſpade; and the ſides of the trenches will then be leſs liable to give way than they would, in the ſpring, when the ſoil is filled like a ſponge with water; which ouzing out, from beneath the beds, into the new-made trenches, their ſides become undermined; and can never, afterwards, be made to ſtand properly: on the contrary, if the trenches be opened in autumn, and the mold which comes out of them be uſed in filling up the hollows, and laying the ſurface even and round, the winter's rains will not paſs through the ſoil, but will run off the ſurface, and rather aſſiſt in eſtabliſhing the beds, than in rendering them tender.

In March, the beds being firmly eſtabliſhed, and their ſurfaces in good working order, the ſoil ſhould be thoroughly trenched with the ſpade, and truncheons inſerted.

The ſets ſhould be put in, about two feet from each other, and a potatoe plant may be dibbled into the center of each interſpace. During ſummer, the

the ſurface ſhould be kept clean hoed, and the potatoes earthed up, from time to time.

IN autumn, after the potatoes are taken up, the ſoil ought to be drawn towards the roots of the plants, leaving channels between them to carry off the winter's rains. The enſuing ſpring, the plants muſt be looked over, and ſuch as have failed ſhould be replaced with freſh ſtrong ſets.

AFTER this, little more will be neceſſary than keeping down the taller weeds: if, however, in the courſe of three or four years, the plants do not gain entire poſſeſſion of the ſoil, by overcoming the weeds and graſſineſs, they muſt be cut down to the ſtub, the interſpaces dug, the rubbiſh of the ſurface turned in, and the roots of the plants freed from incumbrances, with the hoe: A ſecond crop of potatoes may be taken, and the former treatment repeated.

THUS far as to the *Coppice:* we will conclude this ſection with ſome obſervations on what is termed the OZIER BED; kept for the particular purpoſe of the BASKET MAKER.

NOTWITHSTANDING the Ozier is uſually planted near water, we have good reaſon to believe it affects a *ſound*, if not a dry ſoil. The places it moſt

moſt delights in are drained moors, and the banks of large rivers, both of which are peculiarly dry ſituations: it has no diſlike, however, to being flooded occaſionally, but ſeems to be invigorated by ſuch irrigation: therefore, the ſand banks, which we frequently ſee thrown up by the ſides of rivers, and which ſometimes lie for half a century before they become profitable, are peculiarly eligible to be converted into Ozier grounds.

THE method of planting an Ozier ground is this: The ſoil being laid perfectly dry, and its ſurface made thoroughly clean, cuttings, of the ſecond or third year's growth, and about twelve inches long, are planted in drills, about two feet and a half aſunder, in the month of March. The cuttings ought to be thruſt ſeven or eight inches deep, leaving four or five inches of head above ground.

THE intervals ſhould be kept ſtirred with a ſmall plow; or, the firſt year, a crop of potatoes may be taken; the drills, in either caſe, muſt be kept perfectly clean with the hand hoe; and, at the approach of winter, the intervals ſhould be ſplit, and the mold thrown to the roots of the young plants, in order to lay them dry and warm, during winter.

In ſpring, it will be well to trim off the firſt year's ſhoots (though not neceſſary), and replace the plants which have failed, with freſh cuttings.

The ſecond ſummer, the intervals muſt be kept ſtirred, the drills hoed, and the plants earthed up, as before, againſt winter.

The enſuing ſpring, the ſtools muſt again be cleared; although the twigs, as yet, will be of little value. But the third cutting they will produce marketable ware, and will increaſe, in quantity and value, until the profits ariſing from them will be almoſt incredible. In ſituations which the Ozier affects, and in countries where the twigs are in demand, Ozier grounds have been known to pay an annual rent of ten pounds an acre! Under ordinary circumſtances, they will, if properly managed, pay four or five.

In Yorkſhire, the "wands" are ſold by the bundle; but in Gloceſterſhire, where Ozier grounds abound, upon the banks of the Severn, the grounds are let, under leaſe, to baſket makers, who keep up the fences, and take upon themſelves the entire management, during the term of the leaſe.

SECTION

SECTION THE FOURTH.

WOODY WASTES.

NO inconsiderable part of the face of this country, taken collectively, is disfigured by lands bearing this description *; the remedy, however, is easy, and the disgrace may soon be removed.

If the soil and situation be favorable to grass or arable produce, grub up the bushes, and clear away the rubbish; but, on the contrary, if the land, either from its own nature, or from the proportion of woodiness which has already got possession of its surface, can be more profitably converted into Woodland, fill up the vacant spaces, in the following manner:

The first business is to fence it round, and the next to cut down the underwood to the stub, and set up the timber trees. If the vacancies be small, they ought to be trenched with the spade; if large, they may be fallowed with the plow; or, in either case,

* See page 119.

caſe, the plants may be put in, without any other preparation, than digging holes to receive them: however, with *this* kind of management, ſucceſs can only be *hoped for*, while under *that* it may be *ſecured*.

THE ſpecies of wood and the mode of propagation depend upon locality, and the ſpecies of plantation required. If underwood be an object, the ſmaller chaſms may be filled up, by layering; for which purpoſe young ſhoots ought to be left, when the bruſh wood is felled, for layers: if timber alone be the deſired object, ſeedling plants may be put in, and acorns or other ſeeds dibbled in the interſpaces: Whether the Wood, the Grove, or the Coppice, be intended, the large ſpaces ought to be filled up in that way; or ſeeds, only, may be ſown in drills, and treated as before directed; or they may be ſcattered in the random manner, and the ſeedlings kept clean by weeding and hand hoeing; or the foſtering care may be left to nature alone: indeed, in *this* kind of way, Woods and Timber Groves may be propagated.

A GENERAL REMARK ON THE PRACTICE OF PLANTING.

WE do not, however, mean to recommend to our readers, here, practices depending on *chance*, after

after having been solicitous to point out those which may be pursued with *certainty*.

Gentlemen, when they set about forming plantations, or raising Woodlands, ought to consider, that the labour, the fencing, the seeds or plants, the rent, and other contingent charges of the land, their own present credit, and their future fame, are *staked*. If, after waiting eight or ten years, a miscarriage take place, the whole is *lost*. On the contrary, if, by judicious methods and careful management, no material failure happen, the prize is *won*; not only the principal but interest is secured: and this by a small additional expence; the trifling difference in labour bestowed upon the after management, only: for the labour in the first instance, rent, &c. &c. &c. are in both cases similar.

A HINT RESPECTING THE MANAGER OF PLANTATIONS.

MUCH depends upon the person to whose care and management plantations are entrusted. If a Gentleman has not leisure, nor inclination, to attend to them himself, he ought to appoint a man of experience; and, if possible, one who is *settled* near the seat of planting; and who is likely to enjoy his appointment for some length of time. For he he who plants ought to expect to nurſe; and having planted he ought to nurſe, becauſe his own credit is at ſtake. On the contrary, a Gentleman who is continually changing his planter, muſt never expect to ſee his plantations ſucceed; for the credit of the preſent riſes upon the miſcarriage of his predeceſſor: he has even an intereſt in neglecting to nurſe; becauſe his own planting will be thereby ſet off to advantage. On the other hand, being without hopes of ſeeing his own labors ſucceed, he loſes a neceſſary ſtimulus: he is not ſufficiently intereſted; having a ready *excuſe*, in the neglect of his ſucceſſor. Theſe are not theoretical deductions, but are drawn from obſervation.

SUBJECT THE SECOND.

RURAL ORNAMENT.

DIVISION THE FIRST.

HISTORY OF THE RURAL ART.

INTRODUCTION.

MANKIND no ſooner find themſelves in faſt poſſeſſion of the *neceſſaries* of life, than they begin to feel a want of its *conveniences*; and theſe obtained, ſeldom fail of indulging in one or more of its various *refinements*. Some men delight in the luxuries of the imagination; others in thoſe of the ſenſes. One man finds his wants ſupplied in the delicacies of the table, while another has recourſe to perfumes and eſſences for relief: few men are inſenſible to the

 gratifications

gratifications of the ear; and men in general are ſuſceptible of thoſe of the eye. The imitative arts of painting and ſculpture have been the ſtudy and delight of civilized nations, in all ages: but the art of embelliſhing Nature, herſelf, has been reſerved for this age, and for this nation!

A FACT the more aſtoniſhing, as ornamented Nature is as much ſuperior to a Painting or a Statue, as a "Reality is to a Repreſentation;" —as the Man himſelf is to his Portrait. That the ſtriking features—the beauties—of Nature, whenever they have been *ſeen*, have always been *admired*, by men of ſenſe and refinement, is undoubtedly true; but why the good offices of art, in ſetting off thoſe features to advantage, ſhould have been ſo long confined to the human perſon alone, is, of all other facts in the Hiſtory of Arts and Sciences, the moſt extraordinary.

THE Tranſlator of D'Ermenonville's Eſſay on Landſcape has attempted to prove, in an introductory diſcourſe, that the art is nothing *new*, for that it was *known* to the Antients, though not *practiſed*. But the evidences, he produces, go no farther than to ſhew, that the Antients were *admirers of Nature in a ſtate of wildneſs*; for, whenever they attempted to *embelliſh* Nature, they appear to have been guided by a kind of Otaheitean taſte;

taste; as the gardens of the Greeks and Romans, like those of modern nations (until of late years in this country), convey to us no other idea, than that of *Nature tatoo'd**.

MR. BURGH, in a Note to his ingenious Commentary upon Mr. Mason's beautiful poem, *The English Garden*, confirms us in these ideas; and, by a quotation from the Younger Pliny, shews the just notions the Antients entertained of the powers of human invention, in associating and polishing the rougher scenes of Nature: for, after giving us a beautiful description of the natural scenery round his Tuscan villa, upon the banks of the Tiber, he acknowledges "the view before him to resemble "a picture beautifully composed, rather than a "work of Nature accidentally delivered."

WE have been told that the English Garden is but a copy of the Gardens of the Chinese: this, however, is founded in Gallic envy rather than in

 truth;

* The inhabitants of Otaheitee, an island in the Southern hemisphere, ornament their bodies by making punctures in the skin with a sharp-pointed instrument, and call it *tatowing*. The African Negroes are still grosser in their ideas of ornament, gashing their cheeks and temples in a manner similar to that practised by the English Butcher in ornamenting a shoulder of mutton, or a Dutch gardener in embellishing the environs of a mansion.

truth; for though their ſtyle of Gardening may not admit of *tatooings* and *topiary works* *, it has as little to do with natural ſcenery as the garden of an antient Roman, or a modern Frenchman:—THE ART OF *aſſiſting* NATURE is, undoubtedly, all our own.

IT cannot fail of proving highly intereſting to our Readers, to trace the riſe of this delightful Art.

MR. WALPOLE, in his *Anecdotes of Painting in England*, has favoured the Public with *A Hiſtory of the modern Taſte in Gardening*. A pen guided by ſo maſterly a hand, muſt ever be productive of information and entertainment, when employed upon a ſubject ſo truly intereſting, as that which is now before us. Deſirous of conveying to our Readers all the information, which we can compreſs with propriety within the limits of our plan, we wiſhed to have given the *ſubſtance* of this valuable paper; but finding it, already, in the language of ſimplicity, and being aware of the miſchiefs which generally enſue in *meddling* with the productions of genius, we had only one alternative; either wholly to tranſcribe, or wholly to reject. *This* we could not do, in ſtrict juſtice to our Readers; for, beſides

* Trees carved by a *Topiarius* into the form of beaſts, birds, &c.

ſides giving us, in detail, the advancement of the art, it throws conſiderable light upon the art itſelf; and being only a ſmall part of a work upon a different ſubject, it is the leſs likely to fall into the hands of thoſe, to whom it cannot fail of proving highly intereſting. We are, therefore, induced to exceed our intended limits, in this reſpect, by making a literal tranſcript; and we have obtained, through the well known liberality of the Author, his permiſſion for ſo doing.

HISTORY OF THE MODERN TASTE IN GARDENING.

'GARDENING was probably one of the firſt 'arts that ſucceeded to that of building houſes, and 'naturally attended property and individual poſſeſ-'ſion. Culinary, and afterwards medicinal herbs 'were the objects of every head of a family: it 'became convenient to have them within reach, 'without ſeeking them at random in woods, in 'meadows, and on mountains, as often as they were 'wanted. When the earth ceaſed to furniſh 'ſpontaneouſly all theſe primitive luxuries, and 'culture became requiſite, ſeparate incloſures for 'rearing herbs grew expedient. Fruits were in 'the ſame predicament, and thoſe moſt in uſe or 'that demanded attention, muſt have entered into

'and extended the domeſtic incloſure. The good
'man Noah, we are told, planted a vineyard,
'drank of the wine, and was drunken, and every
'body knows the conſequences. Thus we ac-
'quired kitchen gardens, orchards, and vineyards.
'I am apprized that the prototype of all theſe
'ſorts was the garden of Eden; but as that Para-
'diſe was a good deal larger than any we read of
'afterwards, being incloſed by the rivers Piſon,
'Gihon, Hiddekel, and Euphrates, as every tree
'that was pleaſant to the ſight and good for food
'grew in it, and as two other trees were likewiſe
'found there, of which not a ſlip or ſucker remains,
'it does not belong to the preſent diſcuſſion.
'After the Fall, no man living was ſuffered to
'enter into the garden; and the poverty and
'neceſſities of our firſt anceſtors hardly allowed
'them time to make improvements in their eſtates
'in imitation of it, ſuppoſing any plan had been
'preſerved. A cottage and a ſlip of ground for
'a cabbage and a gooſeberry buſh, ſuch as we ſee
'by the ſide of a common, were in all probability
'the earlieſt ſeats and gardens: a well and bucket
'ſucceeded to the Piſon and Euphrates. As
'ſettlements increaſed, the orchard and the vine-
'yard followed; and the earlieſt princes of
'tribes poſſeſſed juſt the neceſſaries of a modern
'farmer.

MATTERS,

'MATTERS, we may well believe, remained long 'in this ſituation; and though the generality of 'mankind form their ideas from the import of 'words in their own age, we have no reaſon to 'think that for many centuries the term Garden 'implied more than a kitchen-garden or orchard. 'When a Frenchman reads of the Garden of Eden, 'I do not doubt but he concludes it was ſomething 'approaching to that of Verſailles, with clipt 'hedges, berceaus, and trellis-work. If his de-'votion humbles him ſo far as to allow that, con-'ſidering who deſigned it, there might be a laby-'rinth full of Æſop's Fables, yet he does not con-'ceive, that four of the largeſt rivers in the world 'were half ſo magnificent as an hundred fountains 'full of ſtatues by Girardon. It is thus that the 'word Garden has at all times paſſed for whatever 'was underſtood by that term in different coun-'tries. But that it meant no more than a kitchen-'garden or orchard for ſeveral centuries, is evident 'from thoſe few deſcriptions that are preſerved of 'the moſt famous gardens of antiquity.

'THAT of Alcinous, in the Odyſſey, is the moſt 'renowned in the heroic times. Is there an ad-'mirer of Homer, who can read his deſcription 'without rapture; or who does not form to his 'imagination a ſcene of delights more pictureſque 'than the landſcapes of Tinian or Juan Fernan-

'dez? Yet what was that boaſted Paradiſe with
'which

the Gods ordain'd
To grace Alcinous and his happy land?——POPE.

'Why, diveſted of harmonious Greek and bewitch-
'ing poetry, it was a ſmall orchard and vineyard,
'with ſome beds of herbs, and two fountains, that
'watered them, incloſed within a quickſet hedge.
'The whole compaſs of this pompous garden in-
'cloſed—four acres.

Four acres was th' allotted ſpace of ground,
Fenc'd with a green incloſure all around.

'The trees were apples, figs, pomegranates, pears,
'olives, and vines.

Tall thriving trees confeſs'd the fruitful mold;
The redning apple ripens into gold.
Here the blue fig with luſcious juice o'erflows,
With deeper red the full pomegranate glows.
The branch here bends beneath the weighty pear,
And verdant olives flouriſh round the year.

* * * * * *

Beds of all various herbs, for ever green,
In beauteous order terminate the ſcene.

'Alcinous's garden was planted by the poet, en-
'riched by him with the fairy gift of eternal ſum-
'mer, and, no doubt, an effort of imagination, ſur-
'paſſing any thing he had ever ſeen. As he has
'beſtowed on the ſame happy prince a palace
'with

'with brazen walls and columns of ſilver, he cer-
'tainly intended that the garden ſhould be propor-
'tionably magnificent. We are ſure, therefore,
'that as late as Homer's age, an incloſure of four
'acres, comprehending orchard, vineyard, and
'kitchen garden, was a ſtretch of luxury the world
'at that time had never beheld.

'THE hanging gardens of Babylon were a ſtill
'greater prodigy. We are not acquainted with
'their diſpoſition or contents, but, as they are ſup-
'poſed to have been formed on terraſſes and the
'walls of the palace, whither ſoil was conveyed on
'purpoſe, we are very certain of what they were
'not; I mean they muſt have been trifling, of no
'extent, and a wanton inſtance of expence and
'labour. In other words, they were what ſump-
'tuous gardens have been in all ages till the
'preſent, unnatural, enriched by art, poſſibly with
'fountains, ſtatues, baluſtrades, and ſummer-houſes,
'and were anything but verdant and rural.

'FROM the days of Homer to thoſe of Pliny,
'we have no traces to lead our gueſs to what were
'the gardens of the intervening ages. When Ro-
'man authors, whoſe climate inſtilled a wiſh for
'cool retreats, ſpeak of their enjoyments in that
'kind, they ſigh for grottos, caves, and the refreſh-
'ing hollows of mountains, near irriguous and
'ſhady

'shady founts; or boast of their porticos, walks of
'planes, canals, baths, and breezes from the sea.
'Their gardens are never mentioned as affording
'shade and shelter from the rage of the dog-star.
'Pliny has left us descriptions of two of his villas.
'As he used his Laurentine villa for his winter
'retreat, it is not surprising that the garden makes
'no considerable part of the account. All he says
'of it is, that the *gestatio* or place of exercise,
'which surrounded the garden (the latter conse-
'quently not being very large), was bounded by
'a hedge of box, and where that was perished,
'with rosemary; that there was a walk of vines,
'and that most of the trees were fig and mulberry,
'the soil not being proper for any other sorts.

'ON his Tuscan villa he is more diffuse; the
'garden makes a considerable part of the de-
'scription:—and what was the principal beauty of
'that pleasure ground? Exactly what was the
'admiration of this country about threescore years
'ago; box-trees cut into monsters, animals, let-
'ters, and the names of the master and the artifi-
'cer. In an age when architecture displayed all
'its grandeur, all its purity, and all its taste; when
'arose Vespasian's amphitheatre, the temple of
'Peace, Trajan's forum, Domitian's baths, and
'Adrian's villa, the ruins and vestiges of which
'still excite our astonishment and curiosity; a
'Roman

'Roman consul, a polished emperor's friend, and a 'man of elegant literature and taste, delighted in 'what the mob now scarce admire in a college 'garden. All the ingredients of Pliny's corre- 'sponded exactly with those laid out by London 'and Wise on Dutch principles. He talks of 'slopes, terraces, a wilderness, shrubs methodically 'trimmed, a marble bason, * pipes spouting water, 'a cascade falling into the bason, bay trees, alter- 'nately planted with planes, and a straight walk, from 'whence issued others parted off by hedges of 'box, and apple trees, with obelisks placed be- 'tween every two. There wants nothing but the 'embroidery of a parterre, to make a garden in 'the reign of Trajan serve for a description of one 'in that of King William †. In one passage 'above,

'* The English gardens described by Hentzner in the reign 'of Elizabeth, are exact copies of those of Pliny. In that at 'Whitehall was a sun-dial and jetd'eau, which, on turning a 'cock, spurted out water and sprinkled the spectators. In 'Lord Burleigh's at Theobalds were obelisks, pyramids, and 'circular porticos, with cisterns of lead for bathing. At 'Hampton Court the garden walls were covered with rose- 'mary, a custom, he says, very common in England. At 'Theobalds was a labyrinth also, an ingenuity I shall mention 'presently to have been frequent in that age.

'† Dr. Plot, in his Natural History of Oxfordshire, p. 380, 'seems to have been a great admirer of trees carved into the 'most heterogeneous forms, which he calls *topiary works*, and 'quotes

'above, Pliny ſeems to have conceived that natu-
'ral irregularity might be a beauty; *in opere urba-*
'*niſſimo*, ſays he, *ſubita velut illati ruris imitatio*.
'Something like a rural view was contrived amidſt
'ſo much poliſhed compoſition. But the idea
'ſoon vaniſhed, lineal walks immediately enve-
'loped the ſlight ſcene, and names and inſcriptions
'in box again ſucceeded to compenſate for the
'daring introduction of nature.

'IN the paintings found at Herculaneum are a
'few traces of gardens, as may be ſeen in the ſecond
'volume of the prints. They are ſmall ſquare in-
'cloſures, formed by trellis work, and eſpaliers *,
'and regularly ornamented with vaſes, fountains,
'and caryatides, elegantly ſymmetrical, and proper
'for the narrow ſpaces allotted to the garden of a
'houſe in a capital city. From ſuch I would not
'baniſh thoſe playful waters that refreſh a ſultry
'manſion

'quotes one Laurembergius for ſaying that the Engliſh are as
'expert as moſt nations in that kind of ſculpture, for which
'Hampton Court was particularly remarkable. The Doctor
'then names other gardens that flouriſhed with animals and
'caſtles, formed *arte topiariâ*, and above all a wren's neſt, that
'was capacious enough to receive a man to ſit on a ſeat made
'within it for that purpoſe.

'* At Warwick Caſtle is an antient ſuit of arras, in which
'there is a garden exactly reſembling theſe pictures of Hercu-
'laneum.

'mansion in town, nor the neat trellis, which pre-
'serves its wooden verdure better than natural
'greens exposed to dust. Those treillages, in the
'gardens at Paris, particularly on the Boulevard,
'have a gay and delightful effect. They form
'light corridores, and transpicuous arbours,
'through which the sun-beams play and chequer
'the shade, set off the statues, vases, and flowers,
'that marry with their gaudy hotels, and suit the
'gallant and idle society who paint the walks be-
'tween their parterres, and realize the fantastic
'scenes of Watteau and Durfé.

'FROM what I have said, it appears how natu-
'rally and insensibly the idea of a kitchen garden
'slid into that which has for so many ages been
'peculiarly termed a Garden, and by our ancestors
'in this country, distinguished by the name of a
'Pleasure garden. A square piece of ground was
'originally parted off in early ages for the use of
'the family:—to exclude cattle, and ascertain the
'property, it was separated from the fields by a
'hedge. As pride, and desire of privacy increased,
'the inclosure was dignified by walls; and, in
'climes where fruits were not lavished by the ri-
'pening glow of nature and soil, fruit-trees were
'assisted and sheltered from surrounding winds by
'the like expedient; for the inundation of luxuries
'which have swelled into general necessities, have
'almost

'almoſt all taken their ſource from the ſimple 'fountain of reaſon.

'WHEN the cuſtom of making ſquare gardens 'incloſed with walls was thus eſtabliſhed, to the 'excluſion of nature and proſpect *, pomp and 'ſolitude combined to call for ſomething that 'might enrich and enliven the inſipid and unani-'mated partition. Fountains, firſt invented for 'uſe, which grandeur loves to diſguiſe and throw 'out of the queſtion, received embelliſhments 'from coſtly marbles, and at laſt, to contradict 'utility, toſſed their waſte of waters into air in 'ſpouting columns. Art, in the hands of rude 'man, had at firſt been made a ſuccedaneum to 'nature; in the hands of oſtentatious wealth, it 'became the means of oppoſing nature; and the 'more it traverſed the march of the latter, the 'more nobility thought its power was demon-'ſtrated. Canals meaſured by the line were 'introduced in lieu of meandering ſtreams, and ter-'races were hoiſted aloft in oppoſition to the facile 'ſlopes that imperceptibly unite the valley to the hill. 'Baluſtrades defended theſe precipitate and dan-'gerous elevations, and flights of ſteps rejoined 'them to the ſubjacent flat from which the terrace

'* It was not uncommon, after the circumadjacent country 'had been ſhut out, to endeavour to recover it by raiſing large 'mounts of earth to peep over the walls of the garden.

'had

'had been dug. Vaſes and ſculpture were added
'to theſe unneceſſary balconies, and ſtatues fur-
'niſhed the lifeleſs ſpot with mimic repreſentations
'of the excluded ſons of men. Thus difficulty
'and expence were the conſtituent parts of thoſe
'ſumptuous and ſelfiſh ſolitudes; and every im-
'provement that was made, was but a ſtep farther
'from nature. The tricks of water-works to wet
'the unwary, not to refreſh the panting ſpectator,
'and parterres embroidered in patterns like a pet-
'ticoat, were but the childiſh endeavours of faſhion
'and novelty to reconcile greatneſs to what it had
'ſurfeited on. To crown theſe impotent diſplays
'of falſe taſte, the ſheers were applied to the
'lovely wildneſs of form with which Nature has
'diſtinguiſhed each various ſpecies of tree and
'ſhrub. The venerable Oak, the romantic
'Beech, the uſeful Elm, even the aſpiring circuit
'of the Lime, the regular round of the Cheſnut,
'and the almoſt moulded Orange Tree, were cor-
'rected by ſuch fantaſtic admirers of ſymmetry.
'The compaſs and ſquare were of more uſe in
'plantations than the nurſery man. The meaſured
'walk, the quincunx, and the etoile, impoſed their
'unſatisfying ſameneſs on every royal and noble
'garden. Trees were headed, and their ſides
'pared away; many French groves ſeem green
'cheſts ſet upon poles. Seats of marble, arbours,
'and ſummer-houſes, terminated every viſto; and
'ſym-

'symmetry, even where the space was too large to
'permit its being remarked at one view, was so
'essential, that, as Pope observed,

——— —each alley has a brother,
And half the garden just reflects the other.

'Knots of flowers were more defensibly subjected
'to the same regularity. Leisure, as Milton ex-
'pressed it,

in trim gardens took his pleasure.

'In the garden of Marshal de Biron at Paris, con-
'sisting of fourteen acres, every walk is buttoned
'on each side by lines of flower-pots, which suc-
'ceed in their seasons. When I saw it, there
'were nine thousand pots of Asters, or la Reine
'Marguerite.

'WE do not precisely know what our ancestors
'meant by a bower; it was probably an arbour;
'sometimes it meant the whole frittered inclosure,
'and in one instance it certainly included a labyrinth.
'Rosamond's bower was indisputably of that kind,
'though whether composed of walls or hedges we
'cannot determine *. A square and a round laby-
'rinth

* 'Drayton in a note to his Epistle of Rosamond, says, her
'labyrinth was built of vaults under ground, arched and
'walled with brick and stone; but, as Mr. Gough observes,
'he

'rinth were ſo capital ingredients of a garden for-
'merly, that in Du Cerceau's architecture, who
'lived in the time of Charles IX. and Henry III.
'there is ſcarce a ground-plot without one of
'each. The enchantment of antique appellations
'has conſecrated a pleaſing idea of a royal reſidence,
'of which we now regret the extinction. Haver-
'ing in the Bower, the jointure of many dow-
'ager queens, conveys to us the notion of a roman-
'tic ſcene.

'IN Kip's Views of the Seats of our Nobility
'and Gentry, we ſee the ſame tireſome and re-
'turning uniformity. Every houſe is approached
'by two or three gardens, conſiſting perhaps of a
'gravel-walk and two graſs-plats, or borders of
'flowers. Each riſes above the other by two or
'three ſteps, and as many walls and terraces, and
'ſo many iron gates, that we recollect thoſe antient
'romances, in which every entrance was guarded
'by nymphs or dragons. At Lady Orford's, at
'Piddletown, in Dorſetſhire, there was, when my
'brother married, a double incloſure of thirteen
'gardens, each I ſuppoſe not much above an hun-
'dred yards ſquare, with an enfilade of correſpon-

'he gives no authority for that aſſertion. V. pref. to 2d edit.
'of Britiſh Topography, p. xxx. Such vaults might remain
'to Drayton's time, but did not prove that there had been no
'ſuperſtructure.'

'dent gates; and before you arrived at these, you 'passed a narrow gut between two stone terraces, 'that rose above your head, and which were 'crowned by a line of pyramidal yews. A bow-'ling-green was all the lawn admitted in those 'times, a circular lake the extent of magnificence.

'YET though these and such preposterous incon-'veniences prevailed from age to age, good sense 'in this country had perceived the want of some-'thing at once more grand and more natural. 'These reflections, and the bounds set to the waste 'made by royal spoilers, gave origin to Parks. 'They were contracted forests, and extended gar-'dens. Hentzner says, that, according to Rous 'of Warwick, the first park was that at Wood-'stock. If so, it might be the foundation of a 'legend that Henry II. secured his mistress in a 'labyrinth: it was no doubt more difficult to find 'her in a park than in a palace, where the intri-'cacy of the woods and various lodges buried in 'covert might conceal her actual habitation.

'IT is more extraordinary that having so long 'ago stumbled on the principle of modern garden-'ing, we should have persisted in retaining its re-'verse, symmetrical and unnatural gardens. That 'parks were rare in other countries, Hentzner, 'who travelled over great part of Europe, leads us

to

'to suppose, by observing that they were com-
'mon in England. In France they retain the
'name, but nothing is more different both in com-
'pass and disposition. Their parks are usually
'square or oblong inclosures, regularly planted
'with walks of chesnuts or limes, and generally
'every large town has one for its public recreation.
'They are exactly like Burton's-court, at Chelsea-
'college, and rarely larger.

'ONE man, one great man we had, on whom
'nor education nor custom could impose their pre-
'judices; who, "on evil days though fallen, and
"with darkness and solitude compassed round,"
'judged that the mistaken and fantastic ornaments
'he had seen in gardens, were unworthy of the
'Almighty Hand that planted the delights of
'Paradise. He seems, with the prophetic eye of
'taste (as I have heard taste well defined*), to
'have conceived, to have foreseen modern garden-
'ing; as Lord Bacon announced the discoveries
'since made by experimental philosophy. The
'description of Eden is a warmer and more just
'picture of the present style than Claud Lorrain
'could have painted from Hagley or Stourhead.

P 2 'The

* 'By the great Lord Chatham who had a good taste
'himself in modern gardening, as he shewed by his own villas
'in Enfield Chace and at Hayes.'

‘ The first lines I shall quote exhibit Stourhead on
‘ a more magnificent scale.

Thro’ Eden went a river large,
Nor chang’d his course, but through the shaggy hill
Pass’d underneath ingulph’d, for God had thrown
That mountain as his garden-mound, high rais’d
Upon the rapid current———

‘ Hagley seems pictured in what follows :

which thro’ veins
Of porous earth with kindly thirst updrawn,
Rose a fresh fountain, and with many a rill
Water’d the garden———

What colouring, what freedom of pencil, what
‘ landscape in these lines !

———from that saphire fount the crisped brooks,
Rolling on orient pearl and sands of gold,
With mazy error under pendent shades
Ran nectar, visiting each plant, and fed
Flow’rs worthy of Paradise, which not *nice art*
In beds and curious knots but *nature* boon
Pour’d forth profuse on hill and dale and plain,
Both where the morning sun first warmly smote
The *open field*, and where the unpierc’d shade
Imbrown’d the noon-tide bow’rs.—*Thus was this place*
A happy rural seat of various view.

‘ Read

'Read this transporting description, paint to your
'mind the scenes that follow, contrast them with
'the savage but respectable terror with which
'the Poet guards the bounds of his Paradise,
'fenced

———with the champaign head
Of a steep wilderness, whose hairy sides
With thicket overgrown, grotesque and wild,
Access denied; and over head upgrew
Insuperable height of loftiest shade,
Cedar and pine, and fir, and branching palm,
A sylvan scene, and as the ranks ascend,
Shade above shade, a woody theatre
Of stateliest view———

'and then recollect that the author of this sublime
'vision had never seen a glimpse of any thing like
'what he has imagined, that his favorite Antients
'had dropped not a hint of such divine scenery,
'and that the conceits in Italian gardens, and
'Theobalds and Nonsuch, were the brightest
'originals that his memory could furnish, His
'intellectual eye saw a nobler plan, so little did he
'suffer by the loss of sight. It sufficed him to
'have seen the materials with which he could
'work. The vigour of a boundless imagination
'told him how a plan might be disposed, that
'would embellish nature, and restore art to its

'proper office, the juſt improvement or imitation 'of it *.

'It is neceſſary that the concurrent teſtimony of 'the age ſhould ſwear to poſterity that the deſcrip'tion above quoted was written above half a cen'tury before the introduction of modern garden'ing, or our incredulous deſcendants will defraud the 'poet of half his glory, by being perſuaded that 'he copied ſome garden or gardens he had ſeen— 'ſo minutely do his ideas correſpond with the 'preſent ſtandard. But what ſhall we ſay for that 'intervening half century who could read that 'plan and never attempt to put it in execution?

'Now let us turn to an admired writer, poſ'terior to Milton, and ſee how cold, how in'ſipid, how taſteleſs is his account of what he 'pronounced a perfect garden. I ſpeak not of 'his ſtyle, which it was not neceſſary for him to 'animate with the colouring and glow of poetry. 'It is his want of ideas, of imagination, of taſte, 'that I cenſure, when he dictated on a ſubject that 'is capable of all the graces that a knowledge of 'beautiful nature can beſtow. Sir William Temple 'was an excellent man; Milton, a genius of the 'firſt order.

'WE

* 'Since the above was written, I have found Milton praiſed 'and Sir William Temple cenſured, on the ſame foundations, 'in a poem called The Riſe and Progreſs of the preſent Taſte 'in Planting, printed in 1767.'

'WE cannot wonder that Sir William declares in 'favour of parterres, fountains, and ſtatues, as 'neceſſary to break the ſameneſs of large graſs-'plats, which he thinks have an ill effect upon the 'eye, when he acknowledges that he diſcovers 'fancy in the gardens of Alcinous. Milton ſtudied 'the Antients with equal enthuſiaſm, but no bigo-'try, and had judgement to diſtinguiſh between the 'want of invention and the beauties of poe-'try. Compare his Paradiſe with Homer's Gar-'den, both aſcribed to a celeſtial deſign. For of Sir 'William, it is juſt to obſerve, that his ideas cen-'tered in a fruit-garden. He had the honour of 'giving to his country many delicate fruits, and he 'thought of little elſe than diſpoſing them to the 'beſt advantage. Here is the paſſage I propoſed 'to quote; it is long, but I need not make an 'apology to the reader for entertaining him with 'any other words inſtead of my own.

"THE beſt figure of a garden is either a ſquare or "an oblong, and either upon a flat or a deſcent: "they have all their beauties, but the beſt I eſteem "an oblong upon a deſcent. The beauty, the air, "the view, makes amends for the expence, which "is very great in finiſhing and ſupporting the "terrace-walks, in levelling the parterres, and in "the ſtone ſtairs that are neceſſary from one to the "other.

"THE perfectest figure of a garden I ever saw, "either at home or abroad, was that of Moor-"park in Hertfordshire, when I knew it about "thirty years ago. It was made by the Countess "of Bedford, esteemed among the greatest wits of "her time, and celebrated by Doctor Donne; "and with very great care, excellent contrivance, "and much cost; but greater sums may be "thrown away without effect or honour, if there "want sense in proportion to money, or *if nature* "*be not followed*, which I take to be the great rule "in this, and perhaps in every thing else, as far as "the conduct not only of our lives but our govern-"ments."

[WE shall see how *natural* that admired garden was.]

"BECAUSE I take * the garden I have named to "have been in all kinds the most beautiful and "perfect, at least in the figure and disposition, that I "have ever seen, I will describe it for a model to "those that meet with such a situation, and are "above the regards of common expence. It lies "on the side of a hill, upon which the house stands, "but

* This garden seems to have been made after the plan laid down by Lord Bacon in his 46th Essay, to which, that I may not multiply quotations, I will refer the reader.

" but not very ſteep. The length of the houſe, " where the beſt rooms and of moſt uſe or plea- " ſure are, lies upon the breadth of the garden; " the great parlour opens into the middle of a " terrace gravel walk that lies even with it, and " which may lie, as I remember, about three " hundred paces long, and broad in proportion; " the border ſet with ſtandard laurels and at large " diſtances, which have the beauty of orange trees " out of flower and fruit. From this walk are " three deſcents by many ſtone ſteps, in the middle " and at each end, into a very large parterre. " This is divided into quarters by gravel-walks, " and adorned with two fountains and eight ſtatues " in the ſeveral quarters. At the end of the " terras-walk are two ſummer-houſes, and the ſides " of the parterre are ranged with two large cloiſters " open to the garden, upon arches of ſtone, and " ending with two other ſummer-houſes even " with the cloiſters, which are paved with ſtone, " and deſigned for walks of ſhade, there being " none other in the whole parterre. Over theſe " two cloiſters are two terraces covered with lead " and fenced with baluſters; and the paſſage into " theſe airy walks is out of the two ſummer-houſes " at the end of the firſt terrace-walk. The cloiſter " facing the ſouth is covered with vines, and " would have been proper for an orange-houſe, " and the other for myrtles or other more com-

" mon

" mon greens, and had, I doubt not, been caſt for
" that purpoſe, if this piece of gardening had been
" then in as much vogue as it is now.

" FROM the middle of this parterre is a deſcent
" by many ſteps flying on each ſide of a grotto
" that lies between them, covered with lead and
" flat, into the lower garden, which is all fruit-trees
" ranged about the ſeveral quarters of a wilder-
" neſs which is very ſhady; the walks here are all
" green, the grotto embelliſhed with figures of
" ſhell rock-work, fountains, and water-works.
" If the hill had not ended with the lower garden,
" and the wall were not bounded by a common
" way that goes through the park, they might have
" added a third quarter of all greens; but this
" want is ſupplied by a garden on the other ſide
" the houſe, which is all of that ſort, very wild,
" ſhady, and adorned with rough rock-work and
" fountains.

" THIS was Moor-park when I was acquainted
" with it, and the ſweeteſt place, I think, that I
" have ſeen in my life, either before or ſince, at
" home or abroad."—

' I WILL make no farther remarks on this de-
' ſcription. Any man might deſign and *build* as
' ſweet a garden, who had been born in and never
' ſtirred

'ſtirred out of Holborn. It was not peculiar to
'Sir William Temple to think in that manner.
'How many Frenchmen are there who have ſeen
'*our* gardens, and ſtill prefer *natural* ſlights of
'ſteps and ſhady cloiſters covered with lead! Le
'Nautre, the architect of the groves and grottos
'at Verſailles, came hither on a miſſion to im-
'prove our taſte. He planted St. James's and
'Greenwich Parks—no great monuments of his
'invention.

'To do farther juſtice to Sir William Temple,
'I muſt not omit what he adds. "What I have
"ſaid of the beſt forms of gardens, is meant only of
"ſuch as are in ſome ſort regular; for there may
"be other forms wholly irregular, that may, for
"aught I know, have more beauty than any of the
"others; but they muſt owe it to ſome extraordi-
"nary diſpoſitions of nature in the ſeat, or ſome
"great race of fancy or judgement in the contri-
"vance, which may reduce many diſagreeing parts
"into ſome figure, which ſhall yet, upon the whole,
"be very agreeable. Something of this I have
"ſeen in ſome places, but heard more of it from
"others, who have lived much among the Chineſes,
"a people whoſe way of thinking ſeems to lie as
"wide of ours in Europe as their country does.
"Their greateſt reach of imagination is employed
"in contriving figures, where the beauty ſhall be
"great

" great and ſtrike the eye, but without any order
" or diſpoſition of parts, that ſhall be commonly or
" eaſily obſerved. And though we have hardly
" any notion of this ſort of beauty, yet they have a
" particular word to expreſs it; and where they
" find it hit their eye at firſt ſight, they ſay the
" Sharawadgi is fine or is admirable, or any ſuch
" expreſſion of eſteem :—but I ſhould hardly adviſe
" any of theſe attempts in the figure of gardens
" among us; they are adventures of too hard at-
" chievement for any common hands; and though
" there may be more honour if they ſucceed well,
" yet there is more diſhonour if they fail, and it is
" twenty to one they will; whereas in regular
" figures, it is hard to make any great and remark-
" able faults."

'FORTUNATELY Kent and a few others were
' not quite ſo timid, or we might ſtill be going up
' and down ſtairs in the open air.

'IT is true, we have heard much lately, as Sir
' William Temple did, of irregularity and imi-
' tations of nature in the gardens or grounds of the
' Chineſe. The former is certainly true: they are
' as whimſically irregular as European gardens are
' formally uniform, and unvaried:—but, with re-
' gard to nature, it ſeems as much avoided, as in
' the ſquares and oblongs, and ſtraight lines of our
' anceſtors.

'ancestors. An artificial perpendicular rock, 'starting out of a flat plain, and connected with 'nothing, often pierced through in various places, 'with oval hollows, has no more pretension to be 'deemed natural, than a lineal terrace, or a parterre. 'The late Mr. Joseph Spence, who had both taste 'and zeal for the present style, was so persuaded of 'the Chinese Emperor's pleasure-ground being laid 'out on principles resembling ours, that he trans-'lated and published, under the name of Sir Harry 'Beaumont, a particular account of that inclosure 'from the Collection of the Letters of the Jesuits. 'I have looked it over, and, except a determined 'irregularity, can find nothing in it that gives me 'any idea of attention being paid to nature. It is 'of vast circumference, and contains 200 palaces, 'besides as many contiguous for the eunuchs, all 'gilt, painted, and varnished. There are raised 'hills from 20 to 60 feet high, streams and lakes, 'and one of the latter five miles round. These 'waters are passed by bridges:—but even their 'bridges must not be straight—they serpentine as 'much as the rivulets, and are sometimes so long as 'to be furnished with resting places, and begin and 'end with triumphal arches. Methinks a straight 'canal is as rational at least as a meandering bridge. 'The colonades undulate in the same manner. In 'short, this pretty gaudy scene is the work of 'caprice and whim, and, when we reflect on their 'buildings,

'buildings, preſents no image but that of unſub
'ſtantial tawdrineſs. Nor is this all. Within this
'fantaſtic Paradiſe is a ſquare town, each ſide a
'mile long. Here the eunuchs of the Court, to
'entertain his Imperial Majeſty with the buſtle and
'buſineſs of the capital in which he reſides, but
'which it is not of his dignity ever to ſee, act mer-
'chants and all ſorts of trades, and even deſignedly
'exerciſe for his royal amuſement every art of
'knavery that is practiſed under his auſpicious go-
'vernment. Methinks, this is the childiſh ſolace
'and repoſe of grandeur, not a retirement from
'affairs to the delights of rural life. Here, too,
'his Majeſty plays at agriculture: there is a quarter
'ſet apart for that purpoſe: the eunuchs ſow, reap,
'and carry in their harveſt, in the imperial pre-
'ſence: and his Majeſty returns to Pekin, per-
'ſuaded that he has been in the country *.

'HAVING

'* The French have, of late years, adopted our ſtyle in
'gardens, but, chuſing to be fundamentally obliged to more
'remote rivals, they deny us half the merit, or rather the ori-
'ginality of the invention, by aſcribing the diſcovery to the
'Chineſe, and by calling our taſte in gardening *le gout*
'*Anglo-Chinois*. I think I have ſhewn that this is a blunder,
'and that the Chineſe have paſſed to one extremity of abſur-
'dity, as the French, and all antiquity, had advanced to the
'other, both being equally remote from nature; regular for-
'mality is the oppoſite point to fantaſtic Sharawadgis. The
'French, indeed, during the faſhionable paroxyſm of philo-
'ſophy,

'HAVING thus cleared my way by aſcertaining
' what have been the ideas on gardening in all
' ages, as far as we have materials to judge by, it
' remains to ſhew to what degree Mr. Kent in-
' vented the new ſtyle, and what hints he had re-
' ceived to ſuggeſt and conduct his undertaking.

' WE

' ſophy, have ſurpaſſed us, at leaſt in meditation on the art.
' I have peruſed a grave treatiſe of recent date, in which the
' author, extending his views beyond mere luxury and amuſe-
' ment, has endeavoured to inſpire his countrymen, even in
' the gratification of their expenſive pleaſures, with benevolent
' projects. He propoſes to them to combine gardening with
' charity, and to make every ſtep of their walks an act of gene-
' roſity, and a leſſon of morality. Inſtead of adorning favou-
' rite points with a heathen temple, a Chineſe pagoda, a Gothic
' tower, or fictitious bridge, he propoſes to them at the firſt
' reſting place to erect a ſchool, a little farther, to found an
' academy, at a third diſtance a manufacture, and, at the ter-
' mination of the park, to endow an hoſpital. Thus, ſays he,
' the proprietor would be led to meditate, as he ſaunters, on
' the different ſtages of human life, and both his expence and
' thoughts would march in a progreſſion of patriotic acts and
' reflections. When he was laying out ſo magnificent, chari-
' table, and philoſophic an Utopian villa, it would have coſt
' no more to have added a Foundling hoſpital, a Senate-houſe,
' and a burying-ground. If I ſmile at ſuch viſions, ſtill one
' muſt be glad, that in the whirl of faſhions, beneficence ſhould
' have its turn in vogue; and though the French treat the
' Virtues like every thing elſe, but as an object of mode, it is
' to be hoped that they too will, every now and then, come
' into faſhion again. The author I have been mentioning
' reminds me of a French Gentleman, who, ſome years ago,

' made

'WE have ſeen what Moor-park was, when
'pronounced a ſtandard. But as no ſucceeding
'generation in an opulent and luxurious country
'contents itſelf with the perfection eſtabliſhed by
'its anceſtors, more perfect perfection was ſtill
'ſought; and improvements had gone on, till
'London and Wiſe had ſtocked our gardens with
'giants, animals, monſters †, coats of arms, and
'mottos, in yew, box, and holly. Abſurdity
'could go no farther, and the tide turned. Bridgman,

'made me a viſit at Strawberry Hill. He was ſo complaiſant
'as to commend the place, and to approve our taſte in gar-
'dens—but in the ſame ſtyle of thinking with the above-cited
'author, he ſaid, " I do not like your imaginary temples and
" fictitious terminations of views: I would have real points of
" view with moving objects; for inſtance, here I would have
" —(I forget what)—and there a watering place." " That
" is not ſo eaſy (I replied); one cannot oblige others to
" aſſemble at ſuch or ſuch a ſpot for one's amuſement—how-
" ever, I am glad you would like a watering-place, for *there*
" happens to be one: in that creek of the Thames, the inha-
" bitants of the village do actually water their horſes: but I
" doubt whether, if it were not *convenient* to them to do ſo,
" they would frequent the ſpot only to enliven my proſpect."—
'Such Gallo-Chinois gardens, I apprehend, will rarely be
'executed.

'† On the piers of a garden gate not far from Paris I ob-
'ſerved two very coquet ſphinxes. Theſe lady monſters had
'ſtraw hats gracefully ſmart on one ſide of their heads, and
'ſilken cloaks half veiling their necks; all executed in ſtone.'

'man, the next fashionable designer of gardens, 'was far more chaste; and whether from good 'sense, or that the Nation had been struck and 'reformed by the admirable paper in The Guar-'dian, No. 173, he banished verdant sculpture, 'and did not even revert to the square precision 'of the foregoing age. He enlarged his plans, 'disdained to make every division tally to its op-'posite, and though he still adhered much to straight 'walks with high clipped hedges, they were only 'his great lines; the rest he diversified by wilder-'ness, and with loose groves of oak, though still 'within surrounding hedges. I have observed in 'the garden * at Gubbins in Hertfordshire many 'detached thoughts, that strongly indicate the 'dawn of modern taste. As his reformation gained 'footing, he ventured farther, and in the royal 'garden at Richmond dared to introduce culti-'vated fields, and even morsels of a forest ap-'pearance, by the sides of those endless and tire-'some walks, that stretched out of one into an-'other without intermission. But this was not till

* 'The seat of the late Sir Jeremy Sambroke. It had 'formerly belonged to Lady More, mother-in-law of Sir 'Thomas More, and had been tyrannically wrenched from 'her by Henry VIII. on the execution of Sir Thomas, though 'not her son, and though her jointure from a former hus-'band.'

'other innovators had broke looſe too from rigid
'ſymmetry.

'BUT the capital ſtroke, the leading ſtep to all
'that has followed, was [I believe the firſt thought
'was Bridgman's] the deſtruction of walls for
'boundaried, and the invention of foſſés—an at-
'tempt then deemed ſo aſtoniſhing, that the com-
'mon people called them Ha! Ha's! to expreſs
'their ſurprize at finding a ſudden and unperceived
'check to their walk.

'ONE of the firſt gardens planted in this ſimple
though ſtill formal ſtyle, was my father's at
'Houghton. It was laid out by Mr. Eyre, an
'imitator of Bridgman. It contains three-and-
'twenty acres, then reckoned a conſiderable
'portion.

'I CALL a ſunk fence the leading ſtep, for theſe
'reaſons. No ſooner was this ſimple enchantment
'made, than levelling, mowing, and rolling, fol-
'lowed. The contiguous ground of the park
'without the ſunk fence was to be harmonized with
'the lawn within; and the garden in its turn was
'to be ſet free from its prim regularity, that it
'might aſſort with the wilder country without.
'The ſunk fence aſcertained the ſpecific garden,
'but that it might not draw too obvious a line of
'diſtinction

'distinction between the neat and the rude, the
'contiguous out-lying parts came to be included
'in a kind of general design; and when nature was
'taken into the plan, under improvements, every
'step that was made, pointed out new beauties and
'inspired new ideas. At that moment appeared
'Kent, painter enough to taste the charms of land-
'scape, bold and opiniative enough to dare and to
'dictate, and born with a genius to strike out a great
'system from the twilight of imperfect essays. He
'leaped the fence, and saw that all nature was a
'garden. He felt the delicious contrast of hill and
'valley changing imperceptibly into each other,
'tasted the beauty of the gentle swell or concave
'scoop, and remarked how loose groves crowned
'an easy eminence with happy ornament, and while
'they called in the distant view between their
'graceful stems, removed and extended the per-
'spective by delusive comparison.

'THUS the pencil of his imagination bestowed
'all the arts of landscape on the scenes he handled.
'The great principles on which he worked were
'perspective, and light and shade. Groupes of
'trees broke too uniform or too extensive a lawn;
'evergreens and woods were opposed to the glare
'of the champaign; and where the view was less
'fortunate, or so much exposed as to be beheld at
'once, he blotted out some parts by thick shades,

 'to

'to divide it into variety, or to make the richeſt
'ſcene more enchanting by reſerving it to a far-
'ther advance of the ſpectator's ſtep. Thus, ſe-
'lecting favourite objects, and veiling deformities
'by ſcreens of plantation; ſometimes allowing the
'rudeſt waſte to add its foil to the richeſt theatre;
'he realized the compoſitions of the greateſt maſ-
'ters in painting. Where objects were wanting to
'animate his horizon, his taſte as an architect could
'beſtow immediate termination. His buildings,
'his ſeats, his temples, were more the works of
'his pencil than of his compaſſes. We owe the
'reſtoration of Greece and the diffuſion of archi-
'tecture to his ſkill in landſcape.

'BUT of all the beauties he added to the face of
'this beautiful country, none ſurpaſſed his manage-
'ment of water. Adieu to canals, circular baſons,
'and caſcades tumbling down marble ſteps, that
'laſt abſurd magnificence of Italian and French
'villas. The forced elevation of cataracts was no
'more. The gentle ſtream was taught to ſer-
'pentize ſeemingly at its pleaſure, and where diſ-
'continued by different levels, its courſe appeared
'to be concealed by thickets properly interſperſed,
'and glittered again at a diſtance where it might
'be ſuppoſed naturally to arrive. Its borders
'were ſmoothed, but preſerved their waving irre-
'gularity. A few trees ſcattered here and there

'on

'on its edges ſprinkled the tame bank that accom-
'panied its meanders; and when it diſappeared
'among the hills, ſhades deſcending from the
'heights leaned towards its progreſs, and framed
'the diſtant point of light under which it was loſt,
'as it turned aſide to either hand of the blue
'horizon.

'THUS, dealing in none but the colours of
'nature, and catching its moſt favourable features,
'men ſaw a new creation opening before their eyes.
'The living landſcape was chaſtened or poliſhed,
'not transformed. Freedom was given to the
'forms of trees; they extended their branches
'unreſtricted, and where any eminent Oak, or
'maſter Beech, had eſcaped maiming and ſurvived
'the foreſt, buſh and bramble was removed, and
'all its honours were reſtored to diſtinguiſh and
'ſhade the plain. Where the united plumage of
'an ancient wood extended wide its undulating
'canopy, and ſtood venerable in its darkneſs, Kent
'thinned the foremoſt ranks, and left but ſo many
'detached and ſcattered trees, as ſoftened the ap-
'proach of gloom, and blended a chequered light
'with the thus lengthened ſhadows of the remain-
'ing columns.

'SUCCEEDING artiſts have added new maſter-
'ſtrokes to theſe touches: perhaps improved or

'brought to perfection ſome that I have named
'The introduction of foreign trees and plants,
'which we owe principally to Archibald Duke of
'Argyle, contributed eſſentially to the richneſs of
'colouring ſo peculiar to our modern landſcape.
'The mixture of various greens, the contraſt of
'forms between our foreſt trees and the northern
'and Weſt Indian firs and pines, are improve-
'ments more recent than Kent, or but little known
'to him. The weeping Willow and every florid
'ſhrub, each tree of delicate or bold leaf, are new
'tints in the compoſition of our gardens. The
'laſt century was certainly acquainted with many
'of thoſe rare plants we now admire. The Wey-
'mouth pine has long been naturalized here; the
'patriarch plant ſtill exiſts at Longleat. The
'light and graceful Acacia was known as early;
'witneſs thoſe ancient ſtems in the court of Bed-
'ford houſe in Bloomſbury-ſquare; and in the
'Biſhop of London's garden at Fulham are many
'exotics of very antient date. I doubt therefore
'whether the difficulty of preſerving them in a
'clime ſo foreign to their nature did not convince
'our anceſtors of their inutility in general; unleſs
'the ſhapelineſs of the lime and horſe cheſnut,
'which accorded ſo well with eſtabliſhed regu-
'larity, and which thence and from their novelty
'grew in faſhion, did not occaſion the neglect of
'the more curious plants.

'BUT

'BUT juſt as the encomiums are that I have be-
'ſtowed on Kent's diſcoveries, he was neither
'without aſſiſtance or faults. Mr. Pope un-
'doubtedly contributed to form his taſte. The
'deſign of the Prince of Wales's garden at Carl-
'ton houſe was evidently borrowed from the Poet's
'at Twickenham. There was a little of affected
'modeſty in the latter, when he ſaid, of all his
'works he was moſt proud of his garden. And
'yet it was a ſingular effort of art and taſte to im-
'preſs ſo much variety and ſcenery on a ſpot of
'five acres. The paſſing through the gloom from
'the grotto to the opening day, the retiring and
'again aſſembling ſhades, the duſky groves, the
'larger lawn, and the ſolemnity of the termination
'at the cypreſſes that lead up to his mother's
'tomb, are managed with exquiſite judgement;
'and though Lord Peterborough aſſiſted him

'To form his quincunx and to rank his vines,

'thoſe were not the moſt pleaſing ingredients of
'his little perſpective.

'I DO not know whether the diſpoſition of the
'garden at Rouſham, laid out for General Dor-
'mer, and in my opinion the moſt engaging of all
'Kent's works, was not planned on the model of
'Mr. Pope's, at leaſt in the opening and retiring
'ſhades of Venus's Vale. The whole is as elegant

'and antique as if the Emperor Julian had ſelected
'the moſt pleaſing ſolitude about Daphne to enjoy
'a philoſophic retirement.

'THAT Kent's ideas were but rarely great, was
'in ſome meaſure owing to the novelty of his art.
'It would have been difficult to have tranſported
'the ſtyle of gardening at once from a few acres to
'tumbling of foreſts: and though new faſhions
'like new religions, [which are new faſhions] often
'lead men to the moſt oppoſite exceſſes, it could
'not be the caſe in gardening, where the experi-
'ments would have been ſo expenſive. Yet it is
'true too that the features in Kent's landſcapes
'were ſeldom majeſtic. His clumps were puny,
'he aimed at immediate effect, and planted not for
'futurity. One ſees no large woods ſketched out
'by his direction. Nor are we yet entirely riſen
'above a too great frequency of ſmall clumps, eſ-
'pecially in the elbows of ſerpentine rivers. How
'common to ſee three or four beeches, then as
'many larches, a third knot of cypreſſes, and a
'revolution of all three! Kent's laſt deſigns were
'in a higher ſtyle, as his ideas opened on ſucceſs.
'The north terrace at Claremont was much ſuperior
'to the reſt of the garden.

'A RETURN of ſome particular thoughts was
'common to him with other painters, and made
'his

'his hand known. A ſmall lake edged by a 'winding bank with ſcattered trees that led to a 'ſeat at the head of the pond, was common to 'Claremont, Eſher, and others of his deſigns. At 'Eſher,

'Where Kent and Nature vied for Pelham's love,

'the proſpects more than aided the Painter's ge-'nius—they marked out the points where his art 'was neceſſary or not; but thence left his judgement 'in poſſeſſion of all its glory.

'HAVING routed profeſſed art, for the modern 'gardener exerts his talents to conceal his art, 'Kent, like other reformers, knew not how to 'ſtop at the juſt limits. He had followed Nature, 'and imitated her ſo happily, that he began to 'think all her works were equally proper for imi-'tation. In Kenſington garden he planted dead 'trees, to give a greater air of truth to the ſcene '—but he was ſoon laughed out of this exceſs. 'His ruling principle was, that Nature abhors a 'ſtraight line. His mimics, for every genius has his 'apes, ſeemed to think that ſhe could love nothing 'but what was crooked. Yet ſo many men of 'taſte of all ranks devoted themſelves to the new 'improvements, that it is ſurprizing how much 'beauty has been ſtruck out, with how few abſur-'dities. Still in ſome lights the reformation ſeems

'to

'to me to have been pushed too far. Though an
'avenue crossing a park or separating a lawn, and
'intercepting views from the seat to which it leads,
'are capital faults, yet a great avenue * cut
'through woods, perhaps before entering a park,
'has a noble air, and

Like footmen running before coaches
To tell the inn what Lord approaches,

'announces the habitation of some man of dis-
'tinction. In other places the total banishment
'of all particular neatness immediately about a
'house, which is frequently left gazing by itself
'in the middle of a park, is a defect. Sheltered
'and even close walks in so very uncertain a cli-
'mate as ours, are comforts ill exchanged for the
'few picturesque days that we enjoy: and when-
'ever a family can purloin a warm and even some-
'thing of an old-fashioned garden from the land-
'scape designed for them by the undertaker in
'fashion, without interfering with the picture, they
'will

'* Of this kind one of the most noble is that of Stanstead,
'the seat of the Earl of Halifax, traversing an antient wood
'for two miles and bounded by the sea. The very extensive
'lawns at that seat, richly inclosed by venerable beech woods,
'and chequered by single beeches of vast size, particularly
'when you stand in the portico of the temple, and survey the
'landscape that wastes itself in rivers of broken sea, recall
'such exact pictures of Claud Lorrain, that it is difficult to
'conceive that he did not paint them from this very spot.'

'will find satisfactions on those days that do not in-
'vite strangers to come and see their improve-
'ments.

'FOUNTAINS have with great reason been
'banished from gardens as unnatural; but it sur-
'prises me that they have not been allotted to their
'proper positions, to cities, towns, and the courts
'of great houses, as proper accompaniments
'to architecture, and as works of grandeur in
'themselves. Their decorations admit the utmost
'invention, and when the waters are thrown up
'to different stages, and tumble over their border,
'nothing has a more imposing or a more refreshing
'sound. A palace demands its external graces
'and attributes, as much as a garden. Fountains
'and cypresses peculiarly become buildings, and
'no man can have been at Rome, and seen
'the vast basons of marble dashed with perpetual
'cascades in the area of St. Peter's, without re-
'taining an idea of taste and splendor. Those in
'the Piazza Navona are as useful as sublimely con-
'ceived.

'GROTTOS in this climate are recesses only to
'be looked at transiently. When they are regu-
'larly composed within of symmetry and architec-
'ture, as in Italy, they are only splendid impro-
'prieties. The most judiciously, indeed most fortunately,

'tunately, placed grotto is that at Stourhead, where
'the river bursts from the urn of its god, and passes
'on its course through the cave.

'BUT it is not my business to lay down rules for
'gardens, but to give the history of them. A
'system of rules pushed to a great degree of refine-
'ment, and collected from the best examples and
'practice, has been lately given in a book intituled,
'Observations on Modern Gardening. The work
'is very ingeniously and carefully executed, and in
'point of utility rather exceeds than omits any
'necessary directions. The author will excuse me
'if I think it a little excess, when he examines that
'rude and unappropriated scene of Matlock-bath,
'and criticises Nature for having bestowed on the
'rapid river Derwent too many cascades. How
'can this censure be brought home to gardening?
'The management of rocks is a province can fall
'to few directors of gardens; still in our distant
'provinces such a guide may be necessary.

'THE author divides his subject into gardens,
'parks, farms, and ridings. I do not mean to
'find fault with this division. Directions are re-
'quisite to each kind, and each has its department
'at many of the great scenes from whence he drew
'his observations. In the historic light, I distin-
'guish them into the garden that connects itself
'with

‘ with a park, into the ornamented farm, and into
‘ the foreſt or ſavage garden. Kent, as I have
‘ ſhewn, invented or eſtabliſhed the firſt ſort.
‘ Mr. Philip Southcote founded the ſecond or *ferme*
‘ *ornee* *, of which is a very juſt deſcription in the
‘ author I have been quoting. The third, I think,
‘ he has not enough diſtinguiſhed. I mean that
‘ kind of alpine ſcene, compoſed almoſt wholly of
‘ pines and firs, a few birch, and ſuch trees as
‘ aſſimilate with a ſavage and mountainous country.
‘ Mr. Charles Hamilton, at Pain’s-hill, in my
‘ opinion, has given a perfect example of this mode
‘ in the utmoſt boundary of his garden. All is
‘ great, and foreign, and rude; the walks ſeem
‘ not deſigned, but cut through the wood of pines;
‘ and the ſtyle of the whole is ſo grand, and con-
‘ ducted with ſo ſerious an air of wild and unculti-
‘ vated extent, that when you look down on this
‘ ſeeming foreſt, you are amazed to find it contain
‘ a very few acres. In general, except as a ſcreen
‘ to conceal ſome deformity, or as a ſhelter in win-
‘ ter, I am not fond of total plantations of ever-
‘ greens. Firs in particular form a very ungraceful
‘ ſummit, all broken into angles.

‘ SIR HENRY ENGLEFIELD was one of the firſt
‘ improvers on the new ſtyle, and ſelected with
‘ ſingular taſte that chief beauty of all gardens,

* At Woburn Farm in Surry.

‘ proſpect

'prospect and fortunate points of view: we tire of
'all the painter's art when it wants these finishing
'touches. The fairest scenes, that depend on
'themselves alone, weary when often seen. The
'Doric portico, the Palladian bridge, the Gothic
'ruin, the Chinese pagoda, that surprize the
'stranger, soon lose their charms to their surfeited
'master. The lake that floats the valley is still
'more lifeless, and its Lord seldom enjoys his ex-
'pence but when he shews it to a visitor. But
'the ornament whose merit soonest fades, is the
'hermitage or scene adapted to contemplation.
'It is almost comic to set aside a quarter of one's
'garden to be melancholy in. Prospect, animated
'prospect, is the theatre that will always be the
'most frequented. Prospects formerly were sacri-
'ficed to convenience and warmth. Thus Bur-
'leigh stands behind a hill, from the top of which
'it would command Stamford. Our ancestors,
'who resided the greatest part of the year at their
'seats, as others did two years together or more,
'had an eye to comfort first, before expence.
'Their vast mansions received and harboured all
'the younger branches, the dowagers and antient
'maiden aunts of the families, and other families
'visited them for a month together. The method
'of living is now totally changed, and yet the same
'superb palaces are still created, becoming a pom-
'pous solitude to the owner, and a transient enter-
'tainment to a few travellers. 'If

'IF any incident aboliſhes or reſtrains the mo-
'dern ſtyle of gardening, it will be this circum-
'ſtance of ſolitarineſs. The greater the ſcene, the
'more diſtant it is probably from the capital, in
'the neighbourhood of which land is too dear to
'admit conſiderable extent of property. Men tire
'of expence that is obvious to few ſpectators.
'Still there is a more imminent danger that
'threatens the preſent, as it has ever done all taſte
'—I mean the purſuit of variety. A modern
'French writer has, in a very affected phraſe, given
'a juſt account of this, I will call it, diſtemper.
'He ſays, *l'ennui du beau amene le gout du ſingulier*.
'The noble ſimplicity of the Auguſtan age was
'driven out by falſe taſte. The gigantic, the
'puerile, the quaint, and at laſt the barbarous and
'the monkiſh, had each their ſucceſſive admirers.
'Muſic has been improved, till it is a ſcience of
'tricks and ſlight of hand: the ſober greatneſs of
'Titian is loſt, and painting, ſince Carlo Maratti,
'has little more relief than Indian paper. Borro-
'mini twiſted * and curled architecture, as if it was
'ſubject to the change of faſhions like a head of hair.
'If we once loſe ſight of the propriety of landſcape
'in our gardens, we ſhall wander into all the fan-
'taſtic Sharawadgis of the Chineſe. We have
'diſcovered the point of perfection. We have

* In particular, he inverted the volutes of the Ionic order.

'given

'given the true model of gardening to the world:
'let other countries mimic or corrupt our taſte;
'but let it reign here on its verdant throne, origi-
'nal by its elegant ſimplicity, and proud of no
'other art than that of ſoftening Nature's harſh-
'neſſes, and copying her graceful touch.

'THE ingenious author of the Obſervations on
'Modern Gardening is, I think, too rigid when he
'condemns ſome deceptions, becauſe they have
'been often uſed. If thoſe deceptions, as a
'feigned ſteeple of a diſtant church, or an unreal
'bridge to diſguiſe the termination of water, were
'intended only to ſurprize, they were indeed tricks
'that would not bear repetition; but being in-
'tended to improve the landſcape, are no more to
'be condemned becauſe common, than they would
'be if employed by a painter in the compoſition
'of a picture. Ought one man's garden to be
'deprived of a happy object, becauſe that object
'has been employed by another? The more we
'exact novelty, the ſooner our taſte will be vitiated.
'Situations are everywhere ſo various, that there
'never can be a ſameneſs, while the diſpoſition of
'the ground is ſtudied and followed, and every
'incident of view turned to advantage.

'IN the mean time, how rich, how gay, how
'pictureſque the face of the country! The demo-
'lition

'lition of walls laying open each improvement, 'every journey is made through a ſucceſſion of 'pictures; and even where taſte is wanting in the 'ſpot improved, the general view is embelliſhed 'by variety. If no relapſe to barbariſm, forma-'lity, and ſecluſion is made, what landſcapes will 'dignify every quarter of our iſland, when the 'daily plantations that are making have attained 'venerable maturity! A ſpecimen of what our 'gardens will be, may be ſeen at Petworth, where 'the portion of the park neareſt the houſe has been 'allotted to the modern ſtyle. It is a garden of 'oaks two hundred years old. If there is a fault 'in ſo auguſt a fragment of improved nature, it is, 'that the ſize of the trees are out of all proportion 'to the ſhrubs and accompaniments. In truth, 'ſhrubs ſhould not only be reſerved for particular 'ſpots and home delight, but are paſſed their 'beauty in leſs than twenty years.

'ENOUGH has been done to eſtabliſh ſuch a 'ſchool of landſcape, as cannot be found on the 'reſt of the globe. If we have the ſeeds of a 'Claude or a Gaſpar amongſt us, he muſt come 'forth. If wood, water, groves, vallies, glades, 'can inſpire or poet or painter, this is the country, 'this is the age to produce them. The flocks, 'the herds, that now are admitted into, now graze 'on the borders of our cultivated plains, are ready

'before the painter's eyes, and group themſelves
'to animate his picture. One misfortune in truth
'there is, that throws a difficulty on the artiſt. A
'principal beauty in our gardens is the lawn and
'ſmoothneſs of turf: in a picture it becomes a
'dead and uniform ſpot, incapable of *chiaro ſcuro*,
'and to be broken inſipidly by children, dogs,
'and other unmeaning figures.

'SINCE we have been familiarized to the ſtudy of
'landſcape, we hear leſs of what delighted our
'ſportſmen anceſtors, *a fine open country*. Wilt-
'ſhire, Dorſetſhire, and ſuch ocean-like extents,
'were formerly preferred to the rich blue proſpects
'of Kent, to the Thames-watered views in Berk-
'ſhire, and to the magnificent ſcale of Nature in
'Yorkſhire. An open country is but a canvas on
'which a landſcape might be deſigned.

'IT was fortunate for the country and Mr. Kent,
'that he was ſucceeded by a very able maſter;
'and did living artiſts come within my plan, I
'ſhould be glad to do juſtice to Mr. Brown; but
'he may be a gainer, by being reſerved for ſome
'abler pen.

'IN general, it is probably true, that the poſ-
'ſeſſor, if he has any taſte, muſt be the beſt de-
'ſigner of his own improvements. He ſees his
'ſituation

'situation in all seasons of the year, at all times of
'the day. He knows where beauty will not clash
'with convenience, and observes in his silent walks
'or accidental rides a thousand hints that must
'escape a person who in a few days sketches out a
'pretty picture, but has not had leisure to examine
'the details and relations of every part.

'TRUTH, which, after the opposition given to
'most revolutions, preponderates at last, will pro-
'bably not carry our style of garden into general
'use on the continent. The expence is only
'suited to the opulence of a free country, where
'emulation reigns among many independent par-
'ticulars. The keeping of our grounds is an
'obstacle, as well as the cost of the first formation.
'A flat country, like Holland, is incapable of
'landscape. In France and Italy the nobility do
'not reside much, and make small expence, at
'their villas. I should think the little princes of
'Germany, who spare no profusion on their palaces
'and country houses, most likely to be our imi-
'tators; especially as their country and climate
'bears, in many parts, resemblance to ours. In
'France, and still less in Italy, they could with
'difficulty attain that verdure which the humidity
'of our clime bestows as the ground-work of our
'improvements. As great an obstacle in France
'is the embargo laid on the growth of their trees.

'As, after a certain age, when they would riſe to
'bulk, they are liable to be marked by the crown's
'ſurveyors as royal timber, it is a curioſity to ſee
'an old tree. A landſcape and a crown-ſurveyor
'are incompatible.'

DIVISION

DIVISION THE SECOND.

PRINCIPLES OF THE RURAL ART.

SECTION THE FIRST.

GENERAL PRINCIPLES.

ARTS, merely imitative, have but one principle to work by, the *nature*, or actual ſtate, of the thing to be imitated. In works of deſign and invention, another principle takes the lead, which is *taſte*. And in every work, in which mental gratification is not the only object, a third principle ariſes, *utility*, or the concurrent purpoſe for which the production is intended.

THE RURAL ART is ſubject to theſe three principles: to nature, as being an imitative art; to utility, as being productive of objects, which are uſeful, as well as ornamental; and to taſte, in the choice of fit objects to be imitated, and of fit pur-

 poſes

poses to be pursued; as also in the composition of the several objects and ends proposed, so as to produce the degree of gratification and use, best suited to the *place*, and to the *purpose* for which it is about to be ornamented: thus, a Hunting Box and a Summer Villa,—an Ornamented Cottage and a Mansion, require a different *style* of ornament, a different *choice* of objects, a different *taste*. Nor can taste be confined to nature and utility,—the place and the purpose, alone; the object of the Polite Arts is the gratification of the human mind, and the state of refinement, of the mind itself, must be considered. Men's notions vary, not only in different ages, but individually in the same age: what would have gratified mankind, a century ago, in this country, will not please them now; while the Country Squire and the Fine Gentleman of the present day require a different kind of gratification: nevertheless, under these various circumstances, every thing may be *natural*, and every thing adapted to the *place*; the *degree of refinement* constituting the principal difference.

WE do not mean to enter into any argument, about whether a state of rusticity, or a state of refinement, whether the forest, or the city, be the state for which the Author of Nature intended the human species: mankind are now found in every state, and in every stage of savageness, rusticity, civili-

civilization, and refinement; and the particular ſtyle of ornament we wiſh to recommend is, that which is beſt adapted to the ſtate of refinement that now prevails in this country; leaving individuals to vary it, as their own peculiar taſtes may direct.

BEFORE we proceed farther, it may be neceſſary to explain what it is we mean, by *nature*, and *natural*. If, in the idea of *natural ſtate*, we include *ground*, *water*, and *wood*, no ſpot in this iſland can be ſaid to be in a *ſtate of nature*. The *ground*, or the ſurface of the earth, as left by Nature (or the convulſions of Nature), remains, it is true, with but few alterations; yet, even here, (eſpecially among rocks and ſteep acclivities, the nobleſt features in the face of Nature), we frequently find the hand of Art has been at work. Again, though rivers may ſtill run in the channels, or nearly in the channels, into which Nature directed them; yet *waters*, taken generally, have been greatly controuled by human art. And, with reſpect to *wood*, we may venture to ſay, that there is not a tree, perhaps not a buſh, now ſtanding upon the face of the country, which owes its identical ſtate of exiſtence to Nature alone. Wherever cultivation has ſet its foot,—wherever the plow and ſpade have laid fallow the ſoil,—Nature is become extinct; and it is in neglected or leſs cultivated places, in moraſſes and mountains, in forests and parochial wastes, we are to seek for anything *near* a state of Nature;—we mean in this country. And who would look for the standard of taste, who expect to find the lovely mixture of wood and lawn, so delightful to the human eye, in the endless woods of America? We may therefore conclude, that the objects of our imitation are not to be sought for in uncultivated Nature. The inhospitable heaths of Westmoreland may astonish for the moment, may be the pleasing amusement of a summer's day, and agreeable objects in their places; but are they *objects of imitation* under the window of a drawing room? Rather let us turn our eyes to well soiled, well wooded, well cultivated spots, where Nature and Art are happily blended; leaving those who are admirers of Art, merely imitative, to contemplate Nature on canvas; and those who wish for Nature, in a state of total neglect, to take up their residence in the woods of America.

FAR be it from us to rebel against the laws of Nature, or to question, in any wise, the perfection of the Deity. A state of nature, in the eye of Omniscience, is undoubtedly a state of perfection. But, in the littleness of human conception, something is wanted, to bring down natural objects to the level of human comprehension. What object in nature is in a state of *human perfection?* Even

in

in the finest woman, a female critic will discover *faults*: and, in the handsomest horse, a buyer will point out what, in the human eye, appear as *imperfections*. Did ever a landscape painter find a scene, purely natural, which might not have been improved by the hand of Art, or which he did not actually improve by a stroke of his pencil? A striking feature may sometimes be caught, where little addition is wanted; but in a rich picturable view, which will bear to be placed repeatedly under the eye, a portion of *lawn* is requisite *, and, in the wilds of nature, we know of no such thing.

THERE-

* Mr. GRAY, whose letters to Dr. WARTON, describing the natural scenery of the North of England, have been held out as models of their kind, corroborates our idea.

'Just beyond this, opens one of the sweetest landscapes that 'art ever attempted to imitate. The bosom of the mountain 'spreading here into a broad bason, discovers in the midst 'Grasmere Water: its margin is hollowed into small bays, 'with bold eminences, some of rock, some of *soft turf*, that 'half conceal and vary the figure of the little lake they com- 'mand: from the shore a low promontory pushes itself far into 'the water, and on it stands a white village, with the parish- 'church rising in the midst of it: hanging inclosures, corn- 'fields, and *meadows green as emerald*, with their trees, and 'hedges, and cattle, fill up the whole space from the edge of 'the water: and just opposite to you is a large farm-house, at 'the bottom of a steep *smooth lawn*, embosomed in old woods, 'which climb half way up the mountain side, and discover 'above

THEREFORE, our idea of *natural*, is not confined to *neglected* nature, but extends to *cultivated* nature, to nature *touched* by art, and rendered intelligible to human perception: and we venture to recommend, as objects most worthy the study and imitation of the artists, such *passages in nature*, as give the highest degree of gratification to cultivated minds in general: passages like the following—no matter whether produced by *accident* or *design*—no matter whether it occur in a forest or a park—or whether it occupy the corner of a common, or fill up a conspicuous quarter of an ornamental ground:—a lofty wood hanging on a bold ascent; its broken margin flowing negligently over the bosom of the valley, lying broad and bare beneath, and falling gently to the brink of a river, winding gracefully along the base.——We further beg leave to add, in this place, that if a passage like this—especially if the vale be occupied by cattle, and the whole scene enlivened by the presence of the sun, and animated by the fleeting shadows of clouds, sweeping its varied surface—is incapable of conveying a degree of gratification to the mind of

'above them a broken line of crags, that crown the scene. 'Not a single red tile, no flareing Gentleman's house, or garden walls, break in upon the repose of this little unsuspected 'paradise; but all is peace, rusticity, and happy poverty, in 'its neatest, most becoming attire.'

Gray's Letters to Dr. Warton, p. 181.

of any of our readers, we have no hope of entertaining ſuch a mind, in this part of our performance.

SECTION THE SECOND.

THE SITE.

BY *the Site* we mean, not only the *place* itſelf, but likewiſe ſo much of the *ſurrounding country* as may fall immediately within the view, and unite with the near grounds.

IF the place be already ſuited to the ſurrounding country, and to the particular purpoſe for which it is intended, the aſſiſtance of art is not wanted, the buſineſs of the artiſt is precluded. If it be *nearly* in this ſtate, the *touchings* of art are only required. But if the place be greatly deficient, as places in general are, then it is the duty of the artiſt "to ſupply its defects, to correct its faults, and to improve its beauties."

EVERY PLACE conſiſts either of *ground* alone, or of *ground* and *water*, or of *ground* and *wood*, or of *ground*, *water*, and *wood*,

SECTION

SECTION THE THIRD.

GROUND.

BY GROUND is meant, that portion of ſurface, which is included within the place to be improved; whether that ſurface be *ſwamp*, *lawn*, *roughet*, *broken ground*, or *rock*; and whether it be a *hill*, a *valley*, a *plain*, or a compoſition of *ſwells*, *dips*, and *levels*.

MR. GILPIN, in his excellent *Obſervations on the Wye*, *&c.* (page 62) gives a ſublime deſcription of what ground ought to be.—"Nothing," ſays he, "gives ſo juſt an idea of the beautiful ſwellings of ground, as thoſe of water, where it has ſufficient room to undulate and expand. In ground which is compoſed of very refractory materials, you are preſented often with harſh lines, angular inſertions, and diſagreeable abruptneſſes. In water, whether in gentle or in agitated motion, all is eaſy, all is ſoftened into itſelf; and the hills and the vallies play into each other, in a variety of the moſt beautiful forms. In agitated water, abruptneſſes indeed there are, but yet they are ſuch abrupt-neſſes

nesses as, in some part or other, unite properly with the surface around them; and are on the whole peculiarly harmonious. Now, if the ocean in any of these swellings and agitations could be arrested and fixed, it would produce that pleasing variety, which we admire in ground. Hence, it is common to fetch our images from water, and apply them to land: we talk of an undulating line, a playing lawn, and a billowy surface; and give a much stronger and more adequate idea by such imagery, than plain language could possibly present."

THE exertions of art, however, are here inadequate, and the artist ought not to attempt to create a *mountain*, a *valley*, or a *plain*; and should but rarely meddle, even with the smaller inequalities of grounds. The *rock* stands equally above the reach of human art, and to attempt to make or unmake it is absurd. *Roughets* and *broken ground* may generally be reduced to lawn, or hid with wood; and a *swamp* may be drained, or covered with water; while *lawn* may be varied, at pleasure, with wood, and sometimes with water.

SECTION THE FOURTH.

WATER.

THIS is either *ſea, lake, pool, river, rivulet,* or *rill.*

A BROAD *lake* and a copious *river* are too great for human art to cope with: nevertheleſs, the margin, and the bank, may be ornamented, and the ſurface of the water diſcloſed to advantage. *Rivulets* are often in themſelves delightful, and, where broad waters are wanted, may be turned to great advantage by art. Stowe * affords a proof of what may be accompliſhed even with a *rill.* If the baſe of the valley be broad, a lake may be formed; if narrow, a river.

IN countries where natural waters abound, art may improve, but ſhould not attempt to create: but in places naturally deficient in water, the artiſt may frequently call forth the creative powers with ſucceſs.

* STOWE, the ſeat of the MARQUIS OF BUCKINGHAM, in Buckinghamſhire.

ſucceſs. In any ſituation, however, art muſt miſcarry, if Nature has not furniſhed a ſufficient ſupply of materials: *confined ſtagnant pools* are always diſguſting: *ſtews*, indeed, may often be neceſſary; but, like the kitchen garden, they ought not to be *ſeen*.

SECTION THE FIFTH.

WOOD.

OVER this element of the rural art the power of the artiſt is abſolute; he can increaſe or diminiſh at pleaſure: if the place be over-wooded, he can lighten it with lawn, or with water: if too naked, he can ſupply the deficiency by PLANTING.

IN forming ORNAMENTAL PLANTATIONS, two things are to be conſidered, the *ſpecies of plantation*, and the *ſpecies of tree*.

THE different ſpecies of plantation are the *Wood*, the *Grove*, the *Coppice* or *Thicket*, the *Border* or *Skreen*, the *Maſs Clump* or *Tuft*, the *Group*, and the *Single Tree*.

WOODS,

WOODS, Groves, and extenſive Thickets, are more particularly adapted to the ſides of hills, and elevated ſituations: detached Maſſes, Groups, and Single Trees, to the lower grounds. A naked hill gives an idea of bleakneſs; as a valley filled with wood does that of dankneſs. The Shrubery depends more on the given accompaniments, than on its own natural ſituation.

MUCH depends upon the diſpoſition of the ſeveral diſtinct woodineſſes (whether accidental or deſigned) with reſpect to each other; and much alſo on the reſpective outlines, particularly thoſe of the larger kind. The Atmoſphere and the Earth are equally bountiful, in affording the rural artiſt fit ſubjects for ſtudy. The margins of ſeas and lakes give us, in their bays and promontories, an ample choice of outline; while the blue expanſe, ſcattered with ſummer's clouds, diſcovers infinite variety, both of figure and diſpoſition.

IN the choice of trees, four things are obſervable: the *height*, the *form*, the *colour*, and the *uſe*. *This* is more eſſential to a good choice, than may appear at firſt ſight; nothing heightens the idea of ornament, eſpecially in the eye of the owner, more than utility; nor, on the contrary, does any thing tend to throw a damp on the gratification, more than does the worthleſſneſs of the object before us.

Imme-

Immediately under the eye, the gaudy Shrub, and the ornamental though uſeleſs Exotic, may be admitted; but for more diſtant objects, and in leſs embelliſhed ſituations, the Timber tree ought to prevail. We ſhould endeavour to make ſuch a choice, as will gratify the preſent age, and benefit the future.

IN mixing trees, there is, in reſpect of *height*, a general rule: the talleſt ſhould be made to occupy the central parts, deſcending gradually to the margin: but, with reſpect to *colour*, all precept, perhaps, would be vague; the tints ought to be as wild and various as the evening ſky, tinged by the ſetting ſun.

FOR farther remarks on this ſubject, ſee the following MINUTES IN PRACTICE.

SECTION THE SIXTH.

NATURAL ACCOMPANIMENTS.

THE moſt judicious mixture of wood and lawn appears dull and unintereſting, when unaccompanied by animated nature. What ſprightlineſs and elegance are added to the plain, in the playful attitudes and racings of the horſe ;—and how much additional grandeur the vale receives in the ſcattered herd !—How ſtrikingly beautiful the boſom of a hill enlivened by the paſturing flock !—What gaiety is given to park ſcenery, in the airy action of the fawn ;—and how peculiarly delightful the ſequeſtered lawn, while the hare is preſent ! Even the ſquirrel gives a chearfulneſs to the grove : while the plumy tribes diſperſe an agreeable animation through the whole ſcene.

SECTION THE SEVENTH.

FACTITIOUS ACCOMPANIMENTS.

UNDER this head, we arrange *Fences*, *Walks*, *Roads*, *Bridges*, *Seats*, and *Buildings*.

THE FENCE, where the place is large, becomes neceſſary ; yet the eye diſlikes conſtraint. Our ideas of liberty carry us beyond our own ſpecies : the imagination feels a diſlike in ſeeing even the brute creation in a ſtate of confinement. Beſide, a tall a tall fence frequently hides, from the ſight, objects the moſt pleaſing; not only the flocks and herds themſelves, but the ſurface they graze upon. Theſe conſiderations have brought the *unſeen fence* into general uſe.

THIS ſpecies of barrier, it muſt be allowed, incurs a degree of deception, which can ſcarcely be warranted, upon any other occaſion. In this inſtance, however, it is a ſpecies of fraud which we obſerve in nature's practice: how often have we ſeen two diſtinct herds feeding, to appearance, in the ſame extended meadow; until coming abruptly upon a deep-ſunk rivulet, or an unfordable river, we diſcover the deception.

BESIDES the *ſunk* fence, another ſort of unſeen barrier may be made, though by no means equal to *that*; eſpecially if near the eye. This is conſtructed of paling, painted of the *inviſible green.* If the colour of the back ground were permanent, and that of the paint made exactly to correſpond with it, the deception would, at a diſtance, be complete; but backgrounds, in general, changing with the ſeaſon, this kind of fence is the leſs eligible.

MASSES and Tufts of woodineſs, ſcattered promiſcuouſly on either ſide of an unſeen winding

 fence,

fence, aſſiſt very much in doing away the idea of conſtraint. For by this means

> The wand'ring flocks that brouſe between the ſhades,
> Seem oft to paſs their bounds, the dubious eye
> Decides not if they crop the mead or lawn.
>
> MASON.

THE WALK, in extenſive grounds, is as neceſſary as the Fence. The beauties of the place are diſcloſed that they may be ſeen; and it is the office of the walk to lead the eye from view to view; in order that, while the tone of health is preſerved, by the favourite exerciſe of nature, the mind may be thrown into uniſon, by the harmony of the ſurrounding objects.

THE direction of the walk ought to be guided by the points of view to which it leads, and the nature of the ground it paſſes over: it ought to be made ſubſervient to the natural impediments—the Ground, Wood, and Water—which fall in its way, without appearing to have any direction of its own. It can ſeldom, with propriety, run any diſtance, in a ſtraight line; a thing which rarely occurs in a *natural walk*. The paths of the Negroes, and the Indians, are always crooked; and thoſe of the brute creation are very ſimilar. Mr. Maſon's deſcription of this *Path of Nature* is happily conceived.

> The peaſant driving through each ſhadowy lane
> His team, that bends beneath th' incumbent weight

Of

Of laughing CERES, marks it with his wheel;
At night and morn, the milk-maid's careless step
Has, thro' yon pasture green, from stile to stile
Imprest a kindred curve; the scudding hare
Draws to her dew-sprent seat, o'er thymy heaths,
A path as gently waving.——— *Eng. Gard.* v. 60.

THE ROAD may be a thing of necessity, as an *approach* to the mansion, or a matter of amusement only, as a *drive* or a *ride*, from which the grounds, and the surrounding country, may be seen to advantage. It should be the study of the artist to make the same road answer, as far as may be, the two-fold purpose.

THE Road and the Walk are subject to the same rule of *Nature* and *Use*. The direction ought to be natural and easy, and adapted to the purpose intended. A Road of necessity ought to be straighter than one of mere conveniency: in this, recreation is the predominant idea; in that, utility. But, even in this, the direct line may be dispensed with. The natural roads upon heaths and open downs, and the grassy glades and green roads across forests and extensive wastes, are proper subjects to be studied.

THE BRIDGE should never be seen where it is not wanted: a useless bridge is a deception; deceptions are frauds; and fraud is always hateful;

unleſs when practiſed to avert ſome greater evil. A bridge without water is an abſurdity; and half a one ſtuck up as an eye-trap is a paltry trick, which, though it may ſtrike the ſtranger, cannot fail of diſguſting, when the fraud is found out.

In low ſituations, and wherever water abounds, bridges become *uſeful*, and are therefore *pleaſing objects*: they are looked for, and ought to appear; not as objects of ornament only, but likewiſe as matters of utility. The walk or the road, therefore, ought to be directed in ſuch a manner, as to croſs the water, at the point in which the bridge will appear to the greateſt advantage.

In the conſtruction of bridges, alſo, regard muſt be had to ornament and utility. A bridge is an artificial production, and as ſuch it ought to appear. It ranks among the nobleſt of human inventions: the ſhip and the fortreſs alone excel it. Simplicity and firmneſs are the leading principles in its conſtruction. Mr. Wheatley's obſervation is juſt when he ſays, "The ſingle wooden arch, now much in faſhion, ſeems to me generally miſapplied. Elevated without occaſion ſo much above, it is totally detached from the river; it is often ſeen ſtraddling in the air, without a glimpſe of water to account for it; and the oſtentation of it, as an ornamental object,

object, diverts all that train of ideas, which its uſe, as a communication, might ſuggeſt." (*Obſ. on Mod. Gard.* 73.) But we beg leave to differ from this ingenious Writer when he tells us, that it is "ſpoiled, if adorned; it is disfigured, if only painted of any other than a duſky colour." In a ruſtic ſcene, where Nature wears her own coarſe garb, "the vulgar foot bridge of planks only, guarded on one hand by a common rail, and ſupported by a few ordinary piles," may be in character; but amidſt a diſplay of ornamented Nature, a contrivance of that kind would appear mean and paltry; and would be an affectation of ſimplicity, rather than the lovely attribute itſelf. In cultivated ſcenes, the bridge ought to receive the ornaments which the laws of architectural taſte allow; and the more poliſhed the ſituation, the higher ſhould be the ſtyle and finiſhings.

SEATS have a two-fold uſe; they are uſeful as places of reſt and converſation, and as guides to the *points of view*, in which the beauties of the ſurrounding ſcene are diſcloſed. Every point of view ſhould be marked with a ſeat, and, ſpeaking generally, no ſeat ought to appear, but in ſome favourable point of view. This rule may not be invariable, but it ought ſeldom to be deviated from.

IN the ruder ſcenes of neglected Nature, the ſimple trunk, rough from the woodman's hands, and the butts or ſtools of rooted trees, without any other marks of tools upon them, than thoſe of the ſaw which ſevered them from their ſtems, are ſeats in character; and, in romantic or recluſe ſituations, the cave or the grotto are admiſſible. But whereever human deſign has been executed, upon the natural objects of the place, the ſeat and every other artificial accompaniment ought to be in uniſon; and whether the bench or the alcove be choſen, it ought to be formed and finiſhed, in ſuch a manner, as to unite with the wood, the lawn, and the walk, which lie round it.

THE colour of ſeats ſhould likewiſe be ſuited to ſituations: where uncultivated Nature prevails, the natural brown of the wood itſelf ought not to be altered: but, where the rural art preſides, white, or ſtone colour, has a much better effect.

BUILDINGS may be admitted into ornamented Nature; provided they be at once uſeful and ornamental. Mere ornament without uſe, and mere uſe without ornament, are equally inadmiſſible. Nor ſhould their uſes be diſguiſed; a barn dreſſed up in the habit of a country church, or a farmhouſe figuring away in the fierceneſs of a caſtle, are ridiculous deceptions. A landſcape daubed upon

upon a board, and a painted ſteeple ſtuck up in a wood, are beneath cenſure.

THERE is another ſpecies of uſeleſs ornament, ſtill more offenſive, becauſe more coſtly, than thoſe comparatively innocent eye-traps; we mean TEMPLES. Whether they be dedicated to Bacchus, Venus, Priapus, or any other genius of debauchery, they are, in this age, enlightened with regard to theological and ſcientific knowledge, equally abſurd.

WE are far, however, from wiſhing to exclude architecture from ornamented Nature. We wiſh to ſee it exerciſed, in all its beauty and ſublimity, upon a CHAPEL *, a MAUSOLEUM †, a MONUMENT ‡, —ſcattered judiciouſly among the natural ornaments:

* The late Sir William Harbord, whoſe taſte and judgement, upon every occaſion, diſcovered a goodneſs of heart and a greatneſs of character, has given us a model of this kind, at Gunton, in Norfolk. The pariſh church ſtanding in his park, and being an old unſightly building, he had it taken down, and a *beautiful temple*, under the direction of the Adams', erected upon its ſite.

† The mauſoleum at Caſtle-Howard, in Yorkſhire, the ſeat of the Earl of Carliſle, is a noble building.

‡ The *temple* of Concord and Victory at Stowe, erected to the memory of the great Lord Chatham, is a beautiful *monumental* building.

ments: not too open or conſpicuous, to give them the air of principals; nor too recluſe, to loſe their full effect, as ſubordinate parts of the whole.

IN extenſive grounds, RETREATS, more eſpecially in the remoter parts, are in a degree requiſite; and, if they be *ſeen*, they ought to harmonize with the views in which they appear; and, of courſe, the more poliſhed the ſcene, the more ornamental ſhould be the Retreat,—whether it be the *Room*, the *Portico*, or the more ſimple *Alcove*.

IN ſcenes leſs ornamented, buildings of an economical nature may appear, with good effect. Sir George Warren, at his ſeat near Fetcham in Surrey, has turned a *temple* into a windmill, with great ſucceſs. What was before a uſeleſs pile of maſonry, now ſtands an emblem of activity and induſtry. Under the heads of large artificial lakes, water mills may generally be erected, and with good effect. A corn mill, under proper regulations, and honeſt management, were ever a bleſſing to the poor in its neighbourhood. Subſtantial farm-houſes, and neat comfortable cottages, ſcattered at a proper diſtance, are always pleaſing objects. The retreat and the porter's lodge, being more ſuſceptible of ornament, may be permitted nearer the eye.

DIVISION

DIVISION THE THIRD.

APPLICATION OF THE RURAL ART.

SECTION THE FIRST.

GENERAL APPLICATION.

HAVING thus enumerated the elements, and ſet forth the leading principles of the art, we now proceed to the execution.

WE beg leave to preface this part of our performance with apprizing our Readers, that all which can be written upon this delightful art, muſt be more or leſs general.—All that *ſcience* can do, is to give *a comprehenſive view of the ſubject*; and all that *precept* ſhould attempt, is to lay down *general rules* of practice. The nature of the place itſelf—and the purpoſe for which it is about to be improved, muſt ever determine the particular application.

cation. It follows, that a gentleman who, from long residence, is fully acquainted with the former, and whose will is a rule to the latter, is the properest person to improve his own place;—provided he be intimately acquainted with the *Art*—as well as with the *place* and the *purpose*: the three are equally and essentially necessary to be understood. It would be as great an impropriety, in a gentleman, to set about the execution of a work of this nature, upon a large scale, before he had acquired a comprehensive knowledge of the subject, studied its leading principle from Nature, made ample observation upon places already ornamented, and had established his theory by some actual practice, at least upon a small scale,—as it would be, in a professional artist, to hazard his own reputation, and risque the property of his employer, before he had studied, maturely, the nature of the place, and had been made fully sensible of the intentions of its owner.

THE nature and style of improvement,—*the purpose*,—depends entirely upon the intention and taste of the proprietor, and is, consequently, as various, as the nature of places themselves: nevertheless, improvements in general may be classed under the following heads:

THE HUNTING BOX,

THE ORNAMENTED COTTAGE,

THE

THE VILLA, and
THE PRINCIPAL RESIDENCE.

BUT, before we enter upon the detail, it will be proper to make ſome general obſervations.

IT is unneceſſary to repeat, that wherever Nature, or accident, has already adapted the place to the intended purpoſe, the aſſiſtance of Art is precluded: but wherever Nature is improveable, Art has an undoubted right to ſtep in, and make the requiſite improvement. The diamond, in its natural ſtate, is improveable by art.

IN the lower claſſes of rural improvements, Art ſhould be ſeen as little as may be; and, in the more negligent ſcenes of Nature, every thing ought to appear, as if it had been done by the general laws of Nature, or had grown out of a ſeries of fortuitous circumſtances. But, in the higher departments, Art cannot be hid; and the *appearance* of deſign ought not to be excluded. A human production cannot be made perfectly natural; and, held out as ſuch, it becomes an impoſition. Our art lies in endeavouring to adapt the productions of Nature to human taſte and perception; and, if much art be uſed, let us not *attempt* to hide it. Who conſiders an accompliſhed well dreſſed woman as in a ſtate of Nature? and who, ſeeing a beautiful ground

adorned

adorned with wood and lawn, with water, bridges, and buildings, believes it to be a natural production? Art ſeldom fails to pleaſe when executed in a maſterly manner: nay, it is frequently the deſign and execution, more than the production itſelf, that ſtrikes us. It is the *artifice*, not the *deſign*, which ought to be avoided. It is the *labour*, and not the *art*, which ought to be concealed. A well written poem would be read with leſs pleaſure, if we *knew* the painful exertions it gave riſe to in the compoſition; and the rural artiſt ought, upon every occaſion, to endeavour to avoid labour; or, if indiſpenſably neceſſary, to conceal it. No trace ſhould be left to lead back the mind to the *expenſive toil*. A mound raiſed, a mountain levelled, or a uſeleſs temple built, convey to the mind feelings equally diſguſting.

But though the aids of Art are as eſſential to Rural Ornament, as education is to manners; yet Art may do too much: ſhe ought to be conſidered as the handmaid, not as the miſtreſs, of Nature: and whether ſhe be employed in carving a tree into the figure of an animal, or in ſhaping a view into the form of a *picture*, ſhe is equally culpable. The nature of the place is ſacred. Should this tend to *landſcape*, from ſome principal point of view, aſſiſt Nature, and perfect it; provided this can be done without injuring the views from other points.

points. But do not disfigure the natural features of the place;—do not sacrifice its native beauties to the arbitrary laws of landscape painting.

> Great Nature scorns controul; she will not bear
> One beauty foreign to the spot or soil
> She gives thee to adorn: 'Tis thine alone
> To mend, not change her features. MASON.

IN a picture bounded by its frame, a perfect landscape is looked for: it is of itself a *whole*, and *the frame must be filled*. But it is not so in ornamented Nature: for, if a side-screen be wanting, the eye is not offended with the frame, or the wainscot; but has always some natural, and often pleasing object to receive it. Suppose a room to be hung with one continued rural representation,—would *distinct pictures* be expected? would correct landscapes be looked for? Nature scarcely knows the thing mankind call a *landscape*. The landscape painter seldom, if ever, finds it perfected to his hands;—some addition or alteration is almost always wanted. Every man, who has made his observations upon natural scenery, knows that the Misletoe of the Oak occurs almost as often as a perfectly natural landscape; and to attempt to make up artificial landscape, upon every occasion, is unnatural, and absurd.

IT

IT is far from our intention to intimate any thing the leaſt diſreſpectful to *landſcape painting:* let the ingenious artiſt cull from Nature her choiceſt beauties, and let him aſſociate them, in the manner beſt ſuited to his own ſingle, and permanent point of view: but do not let us carry his production back again to Nature, and contract her unbounded beauties within the limits of a picture frame. If, indeed, the eye were fixed in one point, the trees could be raiſed to their full height at command, and the ſun be made to ſtand ſtill,—the rural artiſt might work by the rules of *light and ſhade,* and compoſe his landſcape by the painter's law. But, while the ſun continues to pour forth its light impartially, and the trees to riſe with ſlow progreſſion, it would be ridiculous to attempt it. Let him rather ſeek out, imitate, and aſſociate, ſuch STRIKING PASSAGES IN NATURE, as are immediately applicable to the place to be improved, without regard to rules of landſcape, merely human;—and let him,

———————— in this and all
Be various, wild, and free, as Nature's ſelf." MASON.

Inſtead of ſacrificing the natural beauties of the place to one formal landſcape, let every ſtep diſcloſe freſh charms unſought for. How ſtrikingly beautiful the changes formed by the iſlands, and their reſpective mountains, in ſailing through the

West

West Indies! The eye does not catch the same view twice: the scene is ever changing, ever delightful.

We should not have offered our sentiments so freely upon landscape, had not a French writer of some eminence *, in a work lately published, laid it down as an invariable rule, that all ornamental grounds should have a complete landscape, to be seen from some part of the house; and to be made from a perspective drawing, previously taken from the window of the saloon, or the top of the mansion. The work, in other respects, has, nevertheless, great merit, and is in fact an ingenious *Essay on English Gardening*. The Frenchman's vanity, however, will not suffer him to make this acknowledgement: no, it is neither ancient, nor modern, nor English, nor Chinese; and there is some reason to suspect, that the Marquis holds out landscape for no other purpose, than to endeavour to give his work the air of originality; for, in other respects, it contains, in effect, what Wheatley and Mason, Kent and Brown, have previously taught and practised.

* The Marquis D'Ermenonville, friend of the celebrated Rousseau, who died at his house, and whose remains were deposited in his grounds, at Ermenonville.

NOTWITHSTANDING, however, the nature of the place ought not to be ſacrificed to the manſion;—the houſe muſt ever be allowed to be a principal in the compoſition. It ought to be conſidered as the center of the ſyſtem; and the rays of art, like thoſe of the ſun, ſhould grow fainter as they recede from the center. The houſe itſelf being entirely a work of art, its immediate environs ſhould be highly finiſhed; but as the diſtance increaſes, the appearance of deſign ſhould gradually diminiſh, until Nature and fortuitouſneſs have full poſſeſſion of the ſcene.

IN general, the approach ſhould be to the back-front, which, in ſuitable ſituations, ought to lie open to the park or paſture grounds. On the ſides more highly ornamented, a well kept gravel walk may embrace the walls; to this the poliſhed lawn and ſhrubery ſucceed; next, the grounds cloſely paſtured; and, laſtly, the ſurrounding country, which ought not to be conſidered as out of the artiſt's reach: for his art conſiſts, not more in decorating particular ſpots, than in *endeavouring* to render the whole face of Nature delightful.

ANOTHER reaſon for this mode of arrangement is, objects immediately under the eye are ſeen more diſtinctly than thoſe at a diſtance, and ought to be ſuch as are pleaſing in the detail. The beauties of a flower a flower can be discerned on a near view, only; while, at a distance, a roughet of coppice wood, and the most elegant arrangement of flowering shrubs, have the same effect. The most rational entertainment, the human mind is capable of receiving, is that of observing the operations of Nature. The foliation of a leaf, the blowing of flowers, and the maturation of fruit, are among the most delightful subjects that a contemplative mind can be employed in. These processes of Nature are slow, and except the object fall spontaneously under the eye of the observer, the inconveniencies of visiting it in a remote part, so far interfere with the more important employments of life, as to blunt, if not destroy, the enjoyment. This is a strong argument in favor of shrubs and flowers being planted under or near our windows, especially those from whence they may be viewed during the hours of leisure and tranquillity.

FURTHER, the vegetable creation being subject to the animal, the shrub may be cropt, or the flower trodden down, in its day of beauty. If, therefore, we wish to converse with Nature in private, intruders must be kept off,—the shrubery be severed from the ground;—yet not in such a manner as to drive away the pasturing stock from our sight. For this reason, the polished lawn ought not to be too extensive, and the fence, which incloses it, should be such, as will not interrupt the view: But whether it be *seen* or *unseen*, *suspected* or *unsuspected*, is a matter of no great import: its utility in protecting the shrubs and flowers,—in keeping the horns of cattle from the window, and the feet of sheep from the gravel and broken ground,—in preserving that neatness on the outside, which ought to correspond with the finishings and furniture within,—render it of sufficient importance to become even a part of the ornament.

BEFORE any step can be taken towards the execution of the design, be it large or small, a map or plan of the place, exactly as it lies in its unimproved state, should be made; with a corresponding sketch, to mark the intended improvements upon. Not a hovel nor a twig should be touched, until the artist has studied maturely the natural abilities of the place, and has decidedly fixed in his mind, and finally settled on his plan, the proposed alterations: and even then, let him "dare with caution."

THERE is a striking similarity between a neglected scene in Nature, and a neglected cottage beauty; and the mode of improvement is, in either instance, similar. If the face unwashed, and uncombed hair, be considered as ornamental,—Art is not wanted. If rustic bloom and native simplicity

plicity be deemed more desirable,—wash the face, and comb the hair in flowing ringlets, and such ornament will be had in its highest perfection. If that elegance of carriage, and gracefulness of deportment, which flow from education and a refined understanding, be thought requisite, Art may be employed in giving this grace and elegance; for thus far she may go with propriety. But, if she do more, she does too much.

IT would be needless to add, that Art may be employed in concealing, or in doing away, the deformities of Nature. But, even in this, she ought to be cautiously circumspect: for, throughout, there is more danger of doing too much, than too little; and nothing should ever be attempted, which cannot be performed in a masterly manner.

SECTION THE SECOND.

HUNTING BOX.

HERE, little is required of Art. Hunting may be called the amusement of Nature; and the place appropriated to it ought to be no farther altered, from its natural state, than decency and conveniency require:—With men who live in the present age of refinement, "a want of decency is a want of ſenſe."

THE ſtyle, throughout, ſhould be *maſculine*. If ſhrubs be required, they ſhould be of the hardier ſorts; the Box, the Holly, the Lauruſtinus. The trees ſhould be the Oak and the Beech, which give, in Autumn, an agreeable variety of foliage, and anticipate, as it were, the ſeaſon of diverſion. A ſuite of paddocks ſhould be ſeen from the houſe; and if a view of diſtant covers can be caught, the background will be compleat. The ſtable, the kennel, and the leaping bar, are the factitious accompaniments; in the conſtruction of which ſimplicity, ſubſtantialneſs, and conveniency, ſhould prevail.

SECTION THE THIRD.

ORNAMENTED COTTAGE.

NEATNESS and ſimplicity ought to mark the ſtyle of this rational retreat. Oſtentation and ſhow ſhould be cautiouſly avoided; even elegance ſhould not be *attempted*; though it may not be *hid*, if it offer itſelf ſpontaneouſly.

NOTHING,

NOTHING, however, ſhould appear vulgar, nor ſhould ſimplicity be pared down to baldneſs; every thing whimſical or expenſive ought to be ſtudiouſly avoided;—chaſteneſs and frugality ſhould appear in every part.

NEAR the houſe, a *ſtudied neatneſs* may take place; but, at a diſtance, *negligence* ſhould rather be the characteriſtic.

IF a taſte for botany lead to a collection of native ſhrubs and flowers, a ſhrubery will be requiſite; but, in this, every thing ſhould be native. A gaudy exotic ought not to be admitted; nor ſhould the lawn be kept cloſe ſhaven; its flowers ſhould be permitted to blow; and the herbage, when mown, ought to be carried off, and applied to ſome uſeful purpoſe.

IN the artificial accompaniments, ornament ſhould be ſubordinate; utility muſt preſide. The buildings, if any appear, ſhould be thoſe in actual uſe in rural economics. If the hovel be wanted, let it appear; and, as a ſide ſcreen, the barn and rick yard are admiſſible; while the dove houſe and poultry yard may enter more freely into the compoſition.

IN fine, the ORNAMENTED COTTAGE ought to exhibit cultivated Nature, in the firſt ſtage of refinement.

finement. It ranks next above the farm houſe. The plain garb of ruſticity may be ſet off to advantage; but the ſtudied ornaments of art ought not to appear. That becoming neatneſs, and thoſe domeſtic conveniencies, which render the rural life agreeable to a cultivated mind, are all that ſhould be aimed at.

SECTION THE FOURTH.

THE VILLA.

HERE, a ſtyle very different from the preceding, ought to prevail: It ought to be *elegant*, *rich*, or *grand*, according to the ſtyle of the houſe itſelf, and the ſtate of the ſurrounding country; the principal buſineſs of the artiſt being to connect theſe two, in ſuch a manner, that the one ſhall not appear naked or ſlareing, nor the other deſolate and inhoſpitable.

If the houſe be ſtately, and the adjacent country rich and highly cultivated, a ſhrubery may intervene, in which Art may ſhew her utmoſt ſkill. Here, the artiſt may even be permitted to *play at landſcape*:

landſcape: for a place of this kind being ſuppoſed to be ſmall, the intention principally ornamental, and the point of view, probably, confined ſimply to the houſe, ſide ſcreens may be formed, and near grounds laid out, ſuitable to the beſt diſtance that can be caught.

IF buildings, or other artificial ornaments, abound in the offscape, ſo as to mark it ſtrongly, they ought alſo to appear, more or leſs, in the near grounds: if the diſtance abound with wood, the near grounds ſhould be thickened, leſt baldneſs ſhould offend; if open and naked, elegance rather than richneſs ought to be ſtudied, leſt heavineſs ſhould appear.

IT is far from being any part of our plan to cavil unneceſſarily at artiſts, whether living or dead; we cannot, however, refrain from expreſſing a concern for the almoſt total neglect of the principles here laid down, in the prevailing practice of a late celebrated artiſt, in ornamenting the vicinages of villas. We mention it the rather, as Mr. BROWN ſeems to have *ſet the faſhion*, and we are ſorry to find it copied by the inferior artiſts of the day. Without any regard to uniting the houſe with the adjacent country, and, indeed, ſeemingly without any regard whatever to the offscape, one invariable plan of embelliſhment prevails; namely, that of

ſtrip-

ſtripping the near ground, entirely naked, or almoſt ſo, and ſurrounding it with a wavy border of ſhrubs, and a gravel walk; leaving the area, whether large or ſmall, one naked ſheet of greenſward.

IN ſmall confined ſpots, this plan may be eligible. We diſlike thoſe bolſtered flower beds, which abound in the ſuburbs of the metropolis, where the broken ground ſometimes exceeds the lawn: nevertheleſs, to our apprehenſion, a ſimple border, round a large unbroken lawn, only ſerves to ſhew what more is wanted. Simplicity in general is pleaſing; but even ſimplicity may be carried to an extreme, ſo as to convey no other idea than that of poverty and baldneſs. Beſides, how often do we ſee in natural ſcenery, the holly and the foxglove flouriſhing at the foot of an oak, and the primroſe and the campion adding charms to the hawthorn, ſcattered over the paſtured lawn? And we conceive that ſingle trees, footed with evergreens and native flowers, and tufts, as well as borders of ſhrubs, are admiſſible in *ornamental*, as well as in *natural* ſcenery.

THE ſpecies of ſhrub ſhould vary with the intention. If the principal intention be a winter retreat, evergreens, and the early-blowing ſhrubs, ſhould predominate; but, in a place to be frequented in ſummer and autumn, the deciduous tribes ought chiefly to be planted.

SECTION

SECTION THE FIFTH.

PRINCIPAL RESIDENCE.

HERE, the whole art centers. The artiſt has, here, full ſcope for a diſplay of taſte and genius. He has an extent of country under his eye, and will endeavour to make the moſt of what nature and accident have ſpread before him.

ROUND a Principal Reſidence, a gentleman may be ſuppoſed to have ſome conſiderable eſtate, and it is not a ſhrubery and a ground only, which fall under the conſideration of the artiſt: he ought to endeavour to diſcloſe to the view, either from the houſe or ſome other point, as much as he conveniently can of the adjacent eſtate. The love of poſſeſſion is deeply planted in every man's breaſt; and places ſhould bow to the gratification of their owners. To curtail the view by an artificial ſide ſcreen, or any other unnatural machinery, ſo as to deprive a man of the ſatisfaction of overlooking his own eſtate, is an abſurdity which no artiſt ought to be permitted to be guilty of. It is very different, however, where the property of another intrudes

trudes upon the eye: Here, the view may, with some colour of propriety, be bounded by a woody ſcreen.

AFTER what has been ſaid under the head GENERAL APPLICATION, little remains to be added, here. Indeed, it would be in vain to attempt to lay down particular rules: different places are marked by ſets of features, as different from each other, as are thoſe of men's faces. Much muſt be left to the ſkill and taſte of the artiſt; and let thoſe be what they may, nothing but mature ſtudy of the natural abilities of the particular place to be improved, can render him equal to the execution, ſo as to make the moſt of the materials that are placed before him.

SOME few general rules may, neverthelefs, be laid down. The approach ought to be conducted in ſuch a manner, that the ſtriking features of the place ſhall burſt upon the view at once: no trick, however, ſhould be made uſe of: all ſhould appear to fall in naturally. In leading towards the houſe, its direction ſhould not be fully in front, nor exactly at an angle, but ſhould paſs obliquely upon the houſe and its accompaniments; ſo that their poſition with reſpect to each other, as well as the perſpective appearance of the houſe itſelf, may vary at every ſtep: and, having ſhewn the front and the prin-

principal wing, or other accompaniment, to advantage, the approach ſhould wind to the back front, which, as has been already obſerved, ought to lie open to the park or paſtured grounds.

THE improvements, and the rooms from which they are to be ſeen, ſhould be in *uniſon.* Thus, the view from the drawing room ſhould be highly embelliſhed, to correſpond with the beauty and elegance within: every thing, here, ſhould be *feminine*—elegant—beautiful—ſuch as attunes the mind to politeneſs and lively converſation. The breakfaſting room ſhould have more maſculine objects in view: wood, water, and an extended country for the eye to roam over; ſuch as allures us, imperceptibly, to the ride or the chace. The eating and banqueting rooms need no *exterior* allurements.

THERE is a harmony in taſte as in muſic: variety, and even wildneſs upon ſome occaſions, may be admitted; but diſcord cannot be allowed. If, therefore, a place be ſo circumſtanced as to conſiſt of properties totally irreconcileable, the parts ought, if poſſible, to be ſeparated in ſuch a manner, that, like the air and the recitative, the adagio and the allegro, in muſic, they may ſet off each other's charms by the contraſt.

DIVISION

DIVISION THE FOURTH.

PRACTICAL REMARKS
ON
ORNAMENTED PLACES.

HAVING attempted, in the foregoing pages, to lay down ſome GENERAL PRINCIPLES of the Rural Art, and having endeavoured to convey ſome general ideas, concerning the APPLICATION of theſe principles, we now proceed to illuſtrate them farther, by ſuch practical remarks as have occurred to us, on examining the different places which have more particularly engaged our attention.

SECTION THE FIRST.

PERSFIELD,

FORMERLY the ſeat of Mr. MORRIS, near Chepſtow, in Monmouthſhire, — a place upon which

which Nature has been peculiarly laviſh of her favors, and which has been ſpoken of, by different writers, in the moſt flattering terms,—was our firſt place of ſtudy.

PERSFIELD is ſituated upon the banks of the river Wye, which divides Glouceſterſhire and Monmouthſhire, and which was formerly the boundary between England and Wales. The general tendency of the river is from North to South; but, about Persfield, it deſcribes, by its winding courſe, the letter S, ſomewhat compreſſed, ſo as to reduce it in length, and increaſe its width. The grounds of Persfield are lifted high above the bed of the river, ſhelving (from the brink of a lofty and ſteep precipice), towards the South Weſt.

THE lower limb of the letter is filled with *Perſe-wood*, which makes a part of Persfield; but is, at preſent, an impenetrable thicket of coppice-wood. This dips to the South Eaſt, down to the water's edge; and, ſeen from the top of the oppoſite rock, has a good effect.

THE upper limb receives the farms of *Llancot*; rich and highly cultivated: broken into incloſures, and ſcattered with groups and ſingle trees: two well looking farm houſes, in the center, and a neat white chapel, on one ſide: altogether, a lovely little

little paradiſaical ſpot. The lowlineſs of its ſituation ſtamps it with an air of meekneſs and humility; and the natural barriers which ſurround it add that of peacefulneſs and ſecurity. Theſe pictureſque farms do not form a low flat bottom, ſubject to be overflowed by the river; but take the form of a gorget, riſing fulleſt in the middle, and falling, on every ſide, gently to the brink of the Wye; except on the Eaſt ſide, where the top of the gorget leans, in an eaſy manner, againſt a range of perpendicular rock; as if to ſhew its diſk, with advantage, to the walks of Persfield.

THIS rock ſtretches acroſs what may be called the Iſthmus, leaving only a narrow paſs down into the fields of Llancot, and joins the principal range of rocks at the lower bend of the river.

To the North, at the head of the letter, ſtands an immenſe rock (or rather a pile of immenſe rocks heaped one upon another), called Windcliff; the top of which is elevated as much above the grounds of Persfield, as thoſe are above the fields of Llancot.

THESE ſeveral rocks, with the wooded precipices on the ſide of Persfield, form a circular incloſure, about a mile in diameter, including Perſe-wood, Llancot, the Wye, and a ſmall meadow, lying at the foot of Windcliff. THE

THE grounds are divided into the upper and lower lawns*, by the approach to the houſe: a ſmall irregular building; ſtanding near the brink of the precipice; but facing down the lower lawn: a beautiful ground, falling 'precipitately every way 'into a valley which ſhelves down in the middle;' and is ſcattered with groups and ſingle trees in an excellent ſtyle.

THE view from the houſe is ſoft, rich, and beautifully picturable:—the lawn and woods of Persfield, and the oppoſite banks of the river:—the Wye, near its mouth, winding through 'meadows green as emerald,' in a manner peculiarly graceful:—the Severn, here very broad, backed by the wooded and highly cultivated hills of Glouceſterſhire, Wiltſhire, and Somerſetſhire. Not one *rock* enters into the compoſition:—The whole view conſiſts of an elegant arrangement of lawn, wood, and water.

THE upper lawn is a leſs beautiful ground, and the view from it, though it command the 'cultivated hills and rich vallies of Monmouthſhire,'

 bounded

* Mr. Wheatley ſays, the park contains about three hundred acres: but we think the two *lawns* cannot contain ſo much; and if the hanging wood at the bottom of the lower lawn, with the face of the Precipice and Perſe-wood be added, they contain a great deal more.

bounded by the Severn, and backed by the Mendip-hills, is much inferior to that from the houſe.

To give variety to the views from Persfield, to diſcloſe the native grandeur which ſurrounds it, and to ſet off its moſt ſtriking features to advantage, walks have been cut through the woods,—and on the face of the precipice,—which border the grounds to the South and Eaſt. The viewer enters theſe walks at the lower corner of the lower lawn.

THE firſt point of view is marked by an alcove, from which are ſeen the bridge and the town of Chepſtow, with its caſtle ſituated, in a remarkable manner, on the very brink of a perpendicular rock, waſhed by the Wye: and, beyond theſe, the Severn ſhews a ſmall portion of its ſilvery ſurface.

PROCEEDING a little farther along the walk, a view is caught, which the pencil might well copy, as a complete landſcape: The caſtle, with the ſerpentine part of the Wye below Chepſtow, *intermixed* in a peculiar manner with the broad waters of the Severn, form the middle ground; which is backed by diſtant hills: the rocks, crowned with wood, lying between the alcove and the caſtle, to the right; and Caſtlehill farm, elevated upon the oppoſite banks of the river, to the left—form the ſide ſkreens. This

This point is not marked, and muſt frequently be loſt to the ſtranger.

THE grotto, ſituated at the head of Perſe-wood, commands a near view of the oppoſite rocks:— magnificent beyond deſcription! The littleneſs of human art was never placed in a more humiliating point of view:—the caſtle of Chepſtow, a *noble fortreſs*, is, compared with theſe natural bulwarks, a mere *houſe of cards*.

ABOVE the grotto, upon the iſthmus of the Perſefield ſide, is a ſhrubery:—ſtrangely miſplaced! an unpardonable intruſion upon the native grandeur of this ſcene. Mr. GILPIN's obſervations upon this—as they are upon moſt occaſions—are juſt. He ſays, 'It is pity the ingenious Embelliſher 'of theſe ſcenes could not have been ſatisfied with 'the great beauties of Nature which he com- 'manded. The ſhruberies he has introduced in 'this part of his improvements, I fear will rather 'be eſteemed paltry.'——'It is not the ſhrub 'which offends: it is the *formal introduction* of it. 'Wild underwood may be an appendage of the 'grandeſt ſcene: it is a beautiful appendage. A 'bed of violets or of lillies may enamel the ground 'with propriety at the foot of an oak; but, if you 'introduce them artificially in a border, you intro- 'duce a trifling formality, and diſgrace the noble

 'object

'object you wish to adorn.'—GILPIN *on the Wye*, p. 42.*)

THE walk now leaves the wood, and opens upon the lower lawn, until coming near the house it enters the alarming precipice facing Llancot; winding along the face of it, in a manner which does great honour to the artist. Sometimes the fragments of rock, which fall in its way, are avoided, at other times, partially removed, so as to conduct the path along a ledge carved out of the rock; and in one instance, a huge fragment, of a somewhat conical shape, and many yards high, is perforated; the path leading through its base. This is a thought which will hand down, to future times, the greatness of Mr. MORRIS's taste: the design and the execution are equally great: not a mark of a tool to be seen; all appears perfectly natural. The archway is made winding, so that, on the approach, it appears to be the mouth of a cave; and, on a nearer view, the idea is strengthened, by an allowable deception; a black dark recess, on the side next the cliff, which, seen from the entrance before the perforation is discovered, appears to be the darksome inlet into the body of the cave.

FROM

* This shrubery was not introduced, as a PLACE OF VIEW; but merely as a pleasure-ground, or flower-garden.

FROM this point, that vaſt incloſure of rocks and precipices, which marks the peculiar magnificence of Persfield, is ſeen to advantage. The area, containing, in this point of view, the fields of Llancot and the lower margin of Perſe-wood, is broken, in a manner peculiarly pictureſque, by the graceful winding of the Wye; here waſhing a low graſſy ſhore, and there ſweeping at the feet of the rocks,—which riſe in ſome places perpendicular from the water: but in general they have a wooded offset at the baſe, above which they riſe to one, two, or perhaps three or four hundred feet high; expoſing their ample fronts, ſilvered by age, and bearded with ivy, growing out of the wrinkle-like ſeams and fiſſures. If one might be allowed to compare the paltry performances of art with the magnificent works of nature, we ſhould ſay, that this incloſure reſembles a prodigious fortreſs, which has lain long in ruins. It is, in reality, one of nature's ſtrongholds; and, as ſuch, has probably been frequently made uſe of.—Acroſs the iſthmus, on the Glouceſterſhire ſide, there are the remains of a deep intrenchment, called to this day the Bulwark; and tradition ſtill teems with the extraordinary warlike feats, that have been performed among this romantic ſcenery,

FROM the perforated rock, the walk leads down to the cold bath (a complete place), ſeated about the midway of the precipice, in this part leſs ſteep: and, from the cold bath, a rough path winds down to the meadow, by the ſide of the Wye, from whence the precipice, on the Persfield ſide, is ſeen with every advantage: the giant fragments, hung with ſhrubs and ivy, riſe in a ghaſtly manner, from among the underwood, and ſhew themſelves in all their native ſavageneſs *.

FROM the cold bath upward, a coach road (very ſteep and difficult) leads to the top of the cliff, at the upper corner of the upper lawn. Near the top of the road, is a point which commands one of the moſt pleaſing views of Persfield. The Wye, ſweeping through a graſſy vale, which opens to the left: Llancot, backed by its rocks, with the Severn immediately behind them, appearing, in this point of view, to be divided from the Wye, by only a ſharp ridge of rock, with a precipice on either ſide: and, behind the Severn, the vale and wooded hills of Glouceſterſhire.

FROM this place, a road leads to the top of Windcliff—aſtoniſhing ſight! The face of nature probably

* There is another way down into this meadow: a kind of winding ſtaircaſe, furrowed out of the face of the precipice, behind the houſe, and leading down into a walk, made on the ſide of the river; but being at preſent out of repair, the deſcent, this way, is rendered very difficult, and ſomewhat dangerous.

bably affords not a more magnificent ſcene! Llancot in all its grandeur; the grounds of Persfield; the caſtle and town of Chepſtow; the graceful windings of the Wye below, and its conflux with the Severn: to the left, the foreſt of Dean: to the right, the rich marſhes and pictureſque mountains of South Wales: a broad view of the Severn, opening its ſea-like mouth; alſo the conflux of the Avon, with merchant ſhips at anchor in King-road, and veſſels of different deſcriptions under ſail: Auſt-Cliff, and the whole vale of Berkeley, backed by the wooded ſwells of Glouceſterſhire; the view terminating in clouds of diſtant hills, riſing one behind another, until the eye becomes unable to diſtinguiſh the earth's billowy ſurface from the clouds themſelves *.

WERE we to ſuggeſt the farther improvement of this place, it would be to ſeparate the *ſublime* from the *beautiful*; ſo that in viewing the one, the eye might not ſo much as ſuſpect that the other was near.

LET the *hanging walk* be conducted entirely along the precipices, or through the thickets, ſo as to dif-

* The waters of the Severn and Wye, being principals in theſe views, and being ſubject to the ebbings and flowings of the tide, which, at the bridge of Chepſtow, riſes to the almoſt incredible height of forty or fifty feet; it follows, that the time of ſpring tide and high water is the propereſt time for going over Persfield.

disclose the natural scenery, without once discovering the lawn, or any other acquired softness. Let the path be as rude as if trodden only by wild beasts and savages, and the resting places, if any, as rustic as possible.

ERASE, entirely, the present shrubery, and lay out another, as elegant as nature and art could render it, before the house, swelling it out into the lawn, towards the stables; between which and the kitchen garden make a narrow winding entrance.

CONVERT the upper lawn into a deer paddock, suffering it to run as wild, rough, and forest-like, as total negligence would render it.

THE viewer would, then, be thus conducted: He would enter the *hanging walk* by a sequestered path, at the lower corner of the lawn*, pursuing it through the wood to beneath the grotto; and round the head land, or winding through Perse-wood, to the perforated rock and the cold bath; without once conceiving an idea (if possible) that art, or at least that much art, had been made use of, in disclosing the natural grandeur of the surrounding objects; which ought to appear as if they presented

* A young plantation, below the entrance into the lower lawn, has been placed as it were for that purpose.

ſented themſelves to his view, or at moſt, as if nothing was wanted, but his own penetration and judgement, to find them out. The walk ſhould, therefore, be conducted in ſuch a manner, that the breaks might be natural, yet the points of view obvious, or requiring nothing but a few blocks or ſtones to mark them. A ſtranger, at leaſt, wants no ſeat here; he is too eager, in the early part of his walk, to think of lounging upon a bench.

FROM the cold bath he would aſcend the ſteep, near the top of which, a commodious bench or benches might be placed: the fatigue of aſcending the hill would require a reſting place; and there are few points, which afford a more pleaſing view than this; it is grand, without being too broad and glaring.

FROM theſe benches he would enter the *foreſt* part. Here the idea of Nature in her primitive ſtate would be ſtrengthened: the roughneſſes and deer to the right, and the rocks in all their native wildneſs to the left. Even Llancot might be ſhut out from the view, by the natural ſhrubery of the cliff. The Lover's Leap, however (a tremendous peep), might remain; but no benches, nor other work of art, ſhould here be ſeen. A natural path, deviating near the brink of the precipice, would bring the viewer down to the lower corner

of

of the park; where benches ſhould be placed in a happy point, ſo as to give a full view of the rocks and native wildneſſes, and, at the ſame time, hide the farm houſes, fields, and other acquired beauties of Llancot.

HAVING ſatiated himſelf with this ſavage ſcene, he would be led, by a ſtill ruſtic path, through the labyrinth—when the ſhrubery, the lawn, with all its appendages, the graceful Wye and the broad ſilver Severn, would break upon the eye, with every advantage of ornamented nature: the tranſition could not fail to ſtrike.

FROM this ſoft ſcene, he would be ſhewn to the top of Windcliff, where, in one view, he would unite the ſublime and beautiful of Persfield.

SECTION THE SECOND.

STOWE,

THE next place we went over, previouſly to the compoſition of the foregoing part of this work, was STOWE, near BUCKINGHAM, the ſeat of EARL TEMPLE, now the MARQUIS OF BUCKINGHAM: a place a place which, on many accounts, claims an early attention *.

STOWE is a creation of art; a contraſt to Persfield. It was among the firſt places which were formed on the principles of modern taſte; and might be ſaid to give birth to the PROFESSION; as being the ſeminary in which the genius of the great profeſſional Artiſt, BROWN, was unfolded.

STOWE is ſituated in a cultivated country, with a ſurface ſomewhat billowy, but without the advantage of bold diſtant views, to give it feature and effect. The ornamented grounds are extenſive; containing, we were told, near four hundred acres; defined by a ſunk fence; and including a dip or ſhallow valley, through which a rill naturally ran.

OUT of theſe ſlender materials; by means of this tame valley, and this trifling ſtream; all that is beautiful at Stowe has been formed: the reſt is planting and maſonry; the mere work of men's hands: facts which prove the excellency of the art of which we are writing; evincing its infant powers in a moſt extraordinary manner.

THE

* This place we ſaw in October 1783.

THE grounds were originally outlined by LORD COBHAM. The lower, or "old part," was laid out by LOVE (to whose history we cannot speak). The upper, or "new part," by BROWN, whose works, we believe, remain as yet the only public records of his history! *

THE old part is marked by a LAKE, or irregular piece of water, of about ten acres in extent; producing in itself, and with the wood on its margin, a pleasing effect; but the cascade, which is occasionally played off from this reservoir, is a trick unworthy of Stowe, and the art to which Stowe owes the beauties it possesses. A waterfall, in a tame site, is unnatural; and the circumstance of waiting until it be let off, renders it almost ridiculous.

THE new part is equally marked by a RIVER, formed with judgment, and good effect; as occupying the lowest ground; winding, naturally, in the bottom of the valley.

THESE

* The above particulars we had from an intelligent guide, who had lived forty years at Stowe, and who shewed the gardens some years for BROWN; adding, that BROWN lived eleven years, as *gardener* and *bailiff* at Stowe: that, during the latter part of his servitude, he had the liberty of laying out grounds for others; that he made the Duke of Grafton's great water, while he lived at Stowe; and that from Stowe he went to Blenheim.

THESE WATERS claim the best attention of the rural artist: they are, indeed, the almost only subjects of study, at Stowe. The PLANTING having been done, at different times, by various hands, and under a varying style of embellishment, has probably undergone much alteration, and has acquired a stiffness of outline, and a heaviness of composition.

THIS heaviness of style is increased by a profusion of BUILDINGS; thrown across each vista, and guarding each glade *. Art has evidently done too much at Stowe. It is over wooded and over built: every thing appears to be sacrificed to Temples; an elegant arrangement of lawn, wood, and water, is seldom to be seen, in open day-light, in these grounds. We recollect but one: this is between the Palladian Bridge and the Gothic Temple, about half way up the rise; where a sweet view of the river, with the lawns and wood on its banks, is caught: but this view being unmarked, it must frequently be passed unnoticed.

WE

* These BUILDINGS, we learnt from the same authority, are all by BROWN; except the *Temple of Venus* by KENT (circumstantial evidence that he had some share in the planting), and except the *Rotundo*, and the *Temple of Bacchus*, by SIR JOHN VANBURGH. *Mr. Walpole*, however, mentions GIBBS, as having had a part in these erections.—*Anecd. of Painting*, Vol. IV. p. 94.

WE do not mean wholly to decry ORNAMENTAL BUILDINGS, in embellished scenery. In places of magnitude, and where the higher degrees of embellishment are required, especially where a magnificent modern-built house forms a principal in the composition, ARCHITECTURAL ORNAMENTS become in a degree requisite. But they should ever appear as *Embellishments* in a scene, and not assume the character of *Principals*; unless, perhaps, in a sequestered part of extensive grounds, where no outlet to the eye, no offscape, can be had; and, there, an ornamental building may appear, as the Principal of an INTERIOR, with advantage. With a view to the study of this subordinate department of the Rural Art, no place, perhaps, is more worthy of the young Artist's attention than STOWE.

IN the higher part of these grounds; near the fluted column, erected, if we recollect rightly, by LADY COBHAM, to the merits of her husband, during his life-time; we were pleased to see some tufts of trees, shrubs, and flowers, growing promiscuously; and in the same natural way, in which we had long wished to see them, in ornamented Nature. These clumps are placed on the edge of the terrace, or unseen fence, which divides the kept grounds from the adjoining fields,—stocked with grazing cattle. They had, of course, a doubly good

good effect; as being in themselves ornamental, and as assisting to mix and assimilate the kept with the unkept grounds. For the latter purpose, however, they were, at the time we saw them, in too high keeping: an error which a little neglect would soon rectify.

To detail the view from every Temple would convey little useful information to our readers. That from the Temple of Concord and Victory (erected, we believe, in honour of the great Lord Chatham) is the most interesting of the interior views. It consists of a narrow grassy valley or dell, thickly wooded, on either side; in a way which we not unfrequently see, in Nature. But the effect is hurt, by two side vistas opening, in a formal manner, upon two obeliscal buildings; from which, in return, the Temple of Concord is seen. This sort of reciprocity of view may often be given with good effect. But it should ever appear as an effect of accident, rather than of design, and cannot please when introduced in a forced or formal manner.

THE eye having dwelt awhile, with pleasure, in this hollow glade, something unnatural in the shelving of the ground was perceived. On closer examination, and still closer enquiry, this beautiful dell was found to be a work of art: not set about, however,

however, with the intent to produce an artificial valley, but an artificial river!

THIS miſcarriage is not brought forward, here, in detriment to the profeſſional character of Mr. BROWN. Every novice, in every art, is liable to commit errors; and one miſtake, in the courſe of an extenſive practice, is but a ſingle blot in writing a volume. We produce it as a leſſon for young artiſts. Water can ſeldom be retained with advantage, in upland ſituations; even where the ſubſtratum is retentive. In places where this is abſorbent, and where the neighbourhood affords no materials to correct the defect, it is in vain to attempt it.

MR. BROWN, however, on diſcovering his error, had great merit in the manner of correcting it. Sloping away the bank of the *river*, and thus forming a *valley*, inſtead of returning the excavated materials to their former ſtate, ſhewed, in a favorable light, his talent for expedient. In the caſe under notice, the effect of the graſſy dell is infinitely better, than any which a weed-grown canal could ever have produced; beſide the injury which water, pent up in that ſituation, muſt have done to the grounds that lie below. A man may diſcover as much talent, in making a retreat, as in gaining a victory.

SECTION

SECTION THE THIRD.

FISHERWICK.

THE Seat of the MARQUIS OF DONEGALL, near LICHFIELD, was the next place which particularly engaged our attention *.

THE natural ſituation of FISHERWICK is ſtill gentler than that of Stowe; where ſome undulation of ſurface gives a degree of variety to the grounds themſelves, and where diſtances, though ſeldom intereſting, are ſometimes caught. But the ſite of Fiſherwick is a flat, without any relief to the eye; except ſome riſing grounds on the banks of the Tame; which, however, though beautiful in themſelves, are not ſeen from Fiſherwick, with advantage; and except a gentle ſwell of ground, which riſes behind the houſe, and which has been judiciouſly choſen as the more immediate ſite of embelliſhment.

AT the foot of this ſwell, ran a conſiderable rivulet, or ſmall brook, ſevering it from the houſe

* In Nov. 1784, and June 1785.

and park: a flat infertile heath; ſuch as we ſee in various parts of this iſland; and ſuch as never fails to diſguſt the eye; more, perhaps, than any other paſſage of ſurface, which the iſland affords.

THE embelliſhments have been effected by breaking the greenſward of the riſing ground, behind the houſe, with planting; the boldeſt and moſt beautiful part of it being judiciouſly preſerved in lawn,—ſcattered with groupes and ſingle trees. The furwer extremity is a continued grove; and the point towards the houſe is alſo planted; to hide the kitchen garden, and to give to this confined ſite, all the feature and expreſſion it was capable of receiving.

IN the dip, between the garden and the park, in which the rivulet formerly ran, a broad REACH OF WATER is formed; winding up to a large and well built ſtone bridge, over which the road from Lichfield paſſes; and its margins are well wooded: circumſtances that unite in giving this Reach of Water, as ſeen from the Grounds, every picturable advantage of a natural River of the firſt magnitude.

IMMEDIATELY below this Reach, an irregular baſon, or lakelet, is formed with the paſſing ſtream. This baſon is open, on one ſide, to the windows; but

but is judiciously backed by planting; and produces a beautiful effect, as seen from the house *.

IN the front of the house, the lawn swells out fully to the park; from which it is separated by a well managed sunk fence. This lawn shelves down, towards the banks of the Tame (deep sunk, unsightly, and unseen, from the grounds of Fisherwick), and embraces the unwooded margin of the lower water. It is naked; except in so far as it is broken by an aged Sycamore in the principal front of the house,—one or more groupes of Planes in the East front,—and an irregular mass of shrubs, well placed upon the brink of the sunk fence, against the park.

THE PARK, containing some five hundred acres, is encircled, in great part, by skreen plantations; on the outside of which is a public road; on the inner side, a chain of Oaks and Elms, placed at such a distance from the paling, as to form a drive round the park; whose flatted surface is broken, and relieved, by large circular clumps; chiefly of

 Scotch

* This effect, however, is, in our opinion, much injured, by a *noisy cascade*, which is formed between these two waters, under the windows of the *library*. A pebbled stream, shaded by Alders, or other Aquatics, would, we think, have been more in character with the site.

Scotch Firs; with ſingle trees interſperſed, to add to the variety.

THE HOUSE is a large and ſplendid pile of building,—in the beſt ſtyle of modern architecture; and, near the ſummit of the ſwell, by the ſide of the lawn, and under the ſhelter of the more diſtant grove, which have been mentioned, ſtands a ſuperb Conſervatory;—a conſpicuous object from the approach, and the only conſpicuous building in the grounds of Fiſherwick *.

THE principal APPROACH is acroſs the park, which it enters at a conſiderable diſtance from the houſe: nevertheleſs, its two open fronts are ſeen from the entrance, and are kept all the way in view from the road; which, however, does not lead in a direct line to the houſe; but bends ſomewhat to the right of it, to give a fuller view of the grounds (which in this line of approach lie to the right of the houſe), as well as of the ſecond or Eaſt front; but arriving near the houſe and grounds, it takes

* A CONSERVATORY, however, though it may appear advantageouſly, as the principal of an interior, is not ornamental, in compoſition. To render it fit, as a receptacle of tender plants, too large a ſurface of glaſs is required, to admit of architectural proportion. We ſpeak of the ſouth front of this ſpecies of building: the north front, being ſuſceptible of ornament, may be rendered picturable in compoſition.

takes a bold ſweep acroſs the principal front, as with the intention of paſſing them entirely; in a direction, however, ſo oblique, as to reach the line of front at the offices, adjoining to the houſe; where, bending ſharply to the right, it enters the lawn, and terminates at a ſuperb portico, in the principal front.

THIS approach, though in the main part it is admirably conducted, has two objectionable things belonging to it. The houſe, as ſeen from the park gate, at more than half a mile diſtance, appears a confuſed maſs of building: not a feature can be diſtinguiſhed: it is ſome time before the eye determines whether it is one or two fronts that are approached. The grounds, too, at that diſtance are indiſcriminate; the whole aſſemblage has the effect of a diſtant proſpect, ſeen from an eminence. Beſide, the unbroken flatneſs, between the houſe and the entrance into the park, offends. Had a few of the maſſes of planting, which are ſcattered over the park, been placed between the lodges and the houſe, the road winding eaſily through them, until it had reached the firſt bend which has been mentioned, the effect would have been better. On leaving the ſkreen of wood to the left, the grounds and houſe would, in that caſe, break upon the eye, in their fulleſt ſplendor and in the beſt point of view.

Its termination is alſo rendered objectionable, by entering the lawn: but, at Fiſherwick, this could not eaſily be avoided: the offices occupy the third front. The artifice of paſſing the principal front, and then returning to it, is the more venial, as ſome elegant pilaſters, placed on the back part of the portico, and apparently moving behind the ſtately columns in front, as they are paſſed, produce a pleaſing effect. Beſide, by this contrivance, the gate of the lawn is brought near the offices, and an unſightly entrance, in the immediate front of the houſe, evaded.

A MERE ſtate entrance may be permitted within barriered grounds. But many are the inconveniences and embarraſſments avoided, by terminating the approach, at an *unguarded* front.

BUT, perhaps, the moſt objectionable part of the operations, at Fiſherwick, is that of encumbering the park with Firs. It may not, however, be too late to ſet about correcting the error. The Scotch Fir, in genial ſituations, is not of long duration; ſoon acquires its higheſt ſtate of profitableneſs; and it might be right, now, to form maſſes of deciduous trees, various in extent and outline, in the interſpaces of the preſent clumps, which may be gradually removed, as they become ripe, and as the deciduous trees may riſe into ſufficient importance, to appear alone.

OF the more highly embellished grounds of Fisherwick, it would be difficult to say too much: even in the recluse parts, we find subjects of instruction. A secluded lawn, in the woody quarter, between the great water and the park, has a charming effect. A lawn amidst extended wood affords the same *relief*, as a mass of wood on a wide expanse of lawn.

THIS woody quarter terminates at the offices; being cut off from the lawn before the house, by the road which leads through the stable yards to the West front; the ordinary approach to the house. And here the walk, which winds through the shrubery, seems to terminate: but before it reaches the extremity, it begins to sink gradually; and, in proceeding, dips beneath an archway, turned under the road; ascending, as leisurely, on the other side of it, until it is raised to the surface of the great lawn. By this admirable expedient, which may frequently be copied with equal advantage, Ladies are enabled to make the entire circuit of the grounds, without setting the foot on a carriage road; except that in the front of the house; which is highly kept, and the materials remarkably good.

THE planting, too, is done in a masterly manner: the trees are well chosen, well arranged, and well planted; are every where luxuriant, and flouriſhing. The Planes and American Firs, which are ſcattered in groups and ſingle trees, over the lawns, and eſpecially over a ſlope ſhelving to the banks of the Tame, are ſuperiorly elegant.

IF we were to cenſure any part of the deſign, with reſpect to planting, it would be, in having crouded the valley or dip, immediately behind the houſe, with foreſt trees; which, with the water, added to the natural flatneſs of the ſituation, will mutually contribute to render the houſe damp, and the air unwholeſome. Yet, foreſeeing the charming effect which lofty groves (ſuch as the preſent plantations will probably become a century hence), embracing the houſe, will certainly have, we admit the propriety of the deſign: and a judicious poſſeſſor will endeavour to prevent the bad, without deſtroying the good, effect.

INDEED, judicious as the interior arrangement and embelliſhment of the grounds of Fiſherwick are in general, they have evidently been ſecondary conſiderations of the artiſt. His great aim has obviouſly been to throw the whole place, as ſeen from the approach, into one grand compoſition; and he has ſucceeded. For in this point of view, the general aſſemblage is not only ſtriking, as a paſſage in ornamented nature; but puts on an air of of magnificence, which Fiſherwick, taken in detail, cannot claim. The park, when croſſed in the direction of the approach, appears extenſive; a ſuite of meadows adjoining to it, and a further ſuite, on the oppoſite ſide of the Tame, encreaſe the apparent extent of the place. The grounds, too, as ſeen from the nearer approach, hanging on the ſwell, and every way endleſs to the eye, contribute to its magnitude and grandeur. But what adds moſt to this idea, and ſhows the talents of the artiſt in the moſt unequivocal light, is a viſta, purpoſely left, between the groves which occupy the extremities of the embelliſhed grounds, with an unſeen fence, which lets in the ſummit of the ſwell, a rich meadow or paſture ground, with the heads of ſome large trees, which appear at different diſtances, beyond it; thus conveying the idea of an extent of rich park lands; or of embelliſhed grounds, in continuation to thoſe which are immediately under the eye. The Houſe, too, ſtately and new, emboſomed in aſpiring groves, and backed by ſome fine old trees that riſe above them,—contributes not a little to make up an ASSEMBLAGE, which gives the eye and the mind great ſatisfaction. BUT THE WHOLE IS BROWN'S. The grounds, the groves, the waters, and the houſe, *are all his own.*

SECTION

SECTION THE FOURTH.

THE LEASOWES.

THE LEASOWES fell next under view *. This place was laid out by its owner, the celebrated poet, SHENSTONE; who lived and died here.

IT is situated in a detached part of Shropſhire, lying between the counties of Worceſter and Stafford. The ſite is more ſtrongly featured than either of the places laſt deſcribed. It occupies the broken ſlope of one of the rugged hills that form a conſiderable proportion of the ſurface of this country; which abounds, for many miles round, with picturable ſcenery: a ſpecies of country frequently found, in the neighbourhood of mountains,—of whoſe ſtyle it partakes, in the general formation of its ſurface; but is on a ſmaller ſcale, and is leſs broken than mountain ſurfaces; being generally covered with productive ſoil; not expoſing bare rocks, or broken precipices.

THE

* In June 1785.

THE houſe is ſeated under the brow of a bold hang that overlooks the place; but upon a riſing ground, which is formed by two narrow dells, that unite below it: thus occupying an elevated ſituation, near the center of the grounds; which it commands, without ſtanding too high and ſtaring.

THE approach is on the lower ſide of the grounds, below the houſe; but there being no inn or accommodation near it; and travelling, with a friend, on horſeback, we left the public road from Birmingham, before we reached the foot of the hill; and, quitting our horſes, entered the grounds, in a more elevated part,—about the midway of the ſlope; thus gaining, at once, ſome general idea of the ſite.

WITH this firſt appearance of the place we were diſappointed. The ground ſeemed leſs broken, and the character of the ſite leſs romantic, than we expected. Indeed, its character, in this line of approach, is beauty: four or five well turned Limes, and an elegant Aſh, riſing on a gentle ſwell, backed by a luxuriant grove of young foreſt trees, welcomed the eye with a modeſt, ſimple, beautiful view.

FALLING in with a made path, we were led down to the lower corner of a paſture ground; where

where a bench marks a wider, and more ſtrongly featured view: the ground uncommonly fine: a beautiful middle ground between two wooded ſkreens; backed by a bold ſteep, alſo hung with wood. A group of Scotch Firs, old and ragged, on the near ground, is a pimple on this fine face,—an honeſt front.

EVERY part of a view, as each article of the ſame dreſs, ſhould be in character: theſe ragged illfavoured Firs, ſtaring on a rugged point, in a rocky, ragged, pictureſque ſcene, would be in place.

FOLLOWING the path, along the bottom of this intereſting incloſure, we entered the larger dingle; a deep rugged gulley, worn by torrents from the hills; ſuch as we ſee in every broken, hilly country: we have croſſed twenty ſuch, as this was by nature, in one morning's diverſion, on the broken margin of the Vale of Pickering: and ſhould probably have croſſed this, unnoticed, had it not been for a dirty little obeliſque, bearing a Latin inſcription, and a few ſeats, which are ſcattered, here and there, in what, if we recollect rightly, is named *Virgil's Grove*. This lettered retreat occupies the bank or brink of the dingle; acroſs which a dam being thrown, a reſervoir of water and a caſcade are at once formed: not a flight of ſteps; but a tolerable

tolerable imitation of a natural waterfall,—of ten or more feet in height; and, below this, a ſhorter fall is ſeen, without a head of water.

UNFORTUNATELY, however, for Art, ſhe could not augment the ſtream; which is naturally much too ſlender, to give full effect to this ingenious device; ſacred, we underſtood, to *Venus*. This caſcade only plays occaſionally; and we arrived at an unintereſting moment.

PURSUING a path, down one ſide of the waterleſs channel, we came to a "root houſe"—a ruſtic alcove; and, paſſing through this, ſoon found ourſelves at the principal entrance: happily deſigned! well calculated to impreſs the mind with romantic imagery; and thoſe who are deſirous of being *enchanted*, at the Leaſowes, ſhould certainly enter here. Indeed, the entire dingle, between this entrance and the reſervoir, is delightful; or would be, if its native rill were permitted to gurgle in its own channel; which, by nature or art, is happily ſtrewed with ſtone and pebbles; and overhung with trees,—that ſtretch their crooked arms, from the high rugged banks which accompany it; filling it with gloom, and an air of ſolitude; which, in contraſt to gayer ſcenery, is ever delightful to minds bending willingly to contemplation.

How

How many paſſages, equally delightful, has Nature furniſhed, in this iſland. All that art can add are ruſtic paths, to render them pervious to human footſteps, with ſuch reſting places as fortuitous circumſtances will ever point out; as the ſhelf of a rock, the trunk of a fallen tree; or natural coves in the banks, furniſhed with blocks or benches: a ſpecies of rural embelliſhment which is procurable at a ſmall expence.

THE path, which accompanies this pebbly channel, leads down to a pool of water, at the junction of the two dells; fed by this and a ſiſter rill, which paſſes occaſionally down the further branch. Over this pool, the church of Hales-owen, backed by a well broken diſtance, produce a picturable view; and near this piece of water, ſtands a mean-looking building, or ruin, or either, named the *Priory*.

PROCEEDING up the dell, by a path which leads towards the houſe, we ſtooped into another ruſtic alcove, formed with the roots of trees, and calked with moſs;—above which appeared another dry caſcade!

WISHING to ſee the economy and effect of one of theſe ingenious contrivances, the perſon who had the care of the grounds obligingly indulged

us;

us; and having examined the reſervoir—a well ſized horſe pool—and ſeen the ſtopple, we took our ſeats in the root houſe, laſt mentioned,—where

"'Twas ſilence all and pleaſing expectation—"

At length, the water guſhed out from among ſome large roots of a tree, falling five or ſix feet perpendicular; preſently we ſaw it again tumbling down another *precipice* (of three or four feet high)—and another; until my companion was in extacy. And having made a graceful ſweep at our feet, it hid its head among ſome roots and well formed rocks. "Very pretty upon my word!" And pray is not the caſcade of tin and horſebeans at Vauxhall very pretty? *Quære*, Did Vauxhall copy after the Leaſowes, or the Leaſowes after Vauxhall?

LEAVING our cool retreat, we climbed the ſteep: an arduous taſk in a hot day. But the views repaid us amply for the toil. The Clent and Wichbury Hills, Kinver Edge, with other picturable eminences, form a variety of pleaſing compoſitions. This natural gallery abounds with lounging places, and long inſcriptions. The path, however, in the part which immediately overlooks the houſe and grounds of the Leaſowes, is well conducted; deviating, naturally, and giving variety of view.

BUT,

But, in the farther part of the ſame ſteep, and lower down the face of it, a ſtraight walk, with a building at one end (a Temple of Pan, or of any other deity or demon), and with a formal viſta in the middle of it, lined out at right angle, in London and Wiſe's beſt manner, appear in a fine hanging grove, which overlooks the paſture ground we firſt entered. This part is probably of more antient date than the reſt.

Finding ourſelves near our horſes, we diſpenſed with the proffered diſplay of the grand caſcade, and ſet out for Hagley; leaving the Leaſowes, notwithſtanding the day was peculiarly favorable to ſhady groves and purling ſtreams, ſomewhat diſappointed. For what is it? An ornamental farm? No ſuch thing. What has farming to do with Temples, Statues, Vaſes, Mottos, Inſcriptions, Mock Priories, and Artificial Caſcades? Yet do away theſe and who would viſit the Leaſowes: for what would it be then? Why, what it is now held out to be;—an ornamental farm; a lovely little ſpot! Let the paths and the benches (or more ſimple ſeats) remain: lay out others acroſs the farm; now ſeemingly much wanted: let the rills babble in rough ſtoney channels (no matter whether altogether natural, or aſſiſted by art); and if a head of water be deemed indiſpenſable, let it be applied to the purpoſe of turning the wheel of a corn a corn mill; a natural appendage to a farm; and ever a pleaſing object in recluſe ſcenery. Had poor Shenſtone adopted this idea, in the outſet of his plan, he might yet (1785) have lived to enjoy his place; or, while he had lived, might have been happy. It was the expenſive baubles we have ſeen, which threw him on the rack of poverty; and probably haſtened the diſſolution of an amiable and valuable man. Strewing pebbles in the channels of the rills, cutting the paths, and providing a few ſimple ſeats; removing the deformities, and ſhewing the natural beauties of the place, and the diſtances it commands, to advantage, would have been a comparatively ſmall expence, which he might have coped with. But does not this view of the Leaſowes ſuggeſt practical ideas? how many places there are, in this iſland, which, with a ſimilar expence, might be rendered equally delightful.

SECTION THE FIFTH.

H A G L E Y.

THIS has long been celebrated as a ſhow place; and is yet in high repute, if we may judge from the concourſe of company and carriages which we found at the inn. A king's plate, or a musίc meeting, could not have created a greater bustle.

HAGLEY is situated only a few miles from the Leasowes; in the same beautifully broken district. The site, like that of the latter, occupies the slope of an extended hill; but the scale is larger, and the features more prominent and striking than those of the Leasowes. The principal feature is a bold headland, or hanging knoll; split by a chasm, down which a slender rill naturally trickled; but which is now interrupted by dams and cascades; and the whole thickly covered with wood, so that no broken ground outwardly appears.

AT the foot of this hanging swell stands the house; from which a still bolder steep is seen, at a short distance; through an open valley or glade; formed by the wooded slope of the first mentioned hill, on one side, and by a shrubery grove, on the other. The house is surrounded by a lawn, of which the glade forms a part; and, below the house and lawn, is an extent of meadow.

THE first view which strikes, at Hagley, is that from the house, up the glade which has been mentioned, and which is scattered with beautiful Esculuses, and margined with fullgrown tufted forest trees, which clothe the slope, and hang down in

loose

loose festoons, at its feet; forming deep and dark recesses. The glade itself, sweeping round a bold festoon of this forest hang, is lost to the eye: which now glances across the public road (sunk low and unperceived) to the face of Wichbury Hill; a sublime passage of ground; a tempest wave of the Bay of Biscay. The part under view is a close bitten sheep walk, scattered with groups and single trees, and terminating with a tall well proportioned obelisk, standing on the summit of the hill. To the right, a grove of Scotch Firs, hanging on an almost perpendicular brow; and, embosomed in these, a sumptuous colonnade is seen. To the left, a lofty wood, which crowns the apparent summit on that part, and closes the view: the most striking composition of ground, wood, and turf, we have ever seen; especially when the glaring white building in the first distance is covered, as it may be, with a handsome tree in the fore-ground.

WHY the obelisk should please so fully in this view, is difficult to account for; but seen, as it is, terminating the view, and upon the summit of the hill, with no other back ground than the clouds, it certainly adds to the general effect:—its colour is that of stone in the quarry; its shape is finely proportioned: it is lightness and elegance itself; perfectly according with the beautiful near-

 grounds;

grounds; which, by the way, are hurt by a ragged, aukward Pear tree, that ought to be removed.

THE church (a low building) which ſtands near the houſe, at the more immediate foot of the ſlope, is inveloped in a deep feſtoon of the foreſt trees that cover this magnificent feature of Hagley.

ABOVE the church yard, is a remarkable congeries of Limes, near ſixty feet high, and fifty feet arm; with a large Wych Elm, twenty-one feet girt; and ſeveral other large old trees.

A RILL prattling in a paved channel, by the ſide of the walk, which leads up to the caſcade, and other interior operations, in this magnificent foreſt ſcenery, is a charming companion in a dry ſultry ſeaſon: unfortunately, too dry for the caſcades of Hagley: the upper ſprings, which feed the reſervoir, being dried up! a circumſtance we ſeriouſly regretted: for, here, the ſite is ſuch, as may be ſuppoſed to produce a natural caſcade; lofty, ſteep, and ſtrongly featured; a wild mountain dingle. ſtrangely disfigured by a poliſhed rotundo, perched near the top of it; mixing in the view, as ſeen from the gapeſee below, with the ſhaggy furniture of this finely ſavage ſcene: which, if farther furniſhed with a mountain torrent, would be at once grand and awful.

IF

IF art muſt needs meddle with natural ſtreams, how much more eligible are irregular falls, than flights of ſteps. In wild, romantic, and eſpecially in rocky ſituations, *Shenſtonian caſcades* may ever produce, momentarily at leaſt, a pleaſing effect. But let them appear in whatever ſituation they may, if a ſufficient ſupply of water cannot be commanded, to feed a perpetual fall, the reflections which follow the idea of playing them off, as raree-ſhows, muſt ever lower the enjoyment.

BESIDE the caſcade, the interior of the wood contains grottos, ſtatues, and fair buildings; but the fairer Oaks with which this magnificent ground may be ſaid to be loaded, and which prove it to have worn its preſent honors for ſome centuries paſt, give the mind the fuller ſatisfaction.

THE views from the top of the park are grand and extenſive; and the wood ſcene, from *Thomſon's Seat,* is nobly fine; but not more ſo than we have ſeen frequently occur, in ſtrongly featured woody countries. The view is much better, in our eye, a little below;—where *Pope's Building* is not ſeen; the ſequeſtered lawn which contains it is enough: a bench is here wanted.

UPON the whole, Hagley, as the Leaſowes, has fallen ſhort of our expectation; which had unfortunately been raiſed too high. The obeliſk ſcene apart, we would not have rode five miles to have ſeen it. The dingle, the wood ſcenes, the ſequeſtered lawns, and the fine timber, are doubtleſs all charming objects; and, to thoſe who have not been in the habits of viewing ſuch ſcenery, are worth going ten times that diſtance to ſee.

INDEED, throughout, there is a greatneſs of taſte, which does the noble artiſt, who embelliſhed it, great honor. It is probable, however, that LORD LYTTELTON was aſſiſted in his deſigns by MR. SHENSTONE, and by other men of taſte and genius, among whom he lived; and often, no doubt, at Hagley.

THE caſcade, and the claſſical alluſions are after the manner of the Leaſowes:—indeed the two places are evidently of the *ſame genus*; their *ſpecific difference* conſiſting in Hagley's being on a larger ſcale, more ſtrongly featured, and more fully wooded. Their embelliſhments, as well as the views from them, are very ſimilar. Their ages, too, are ſimilar: they are both of them growing *ſeedy*. While they flouriſhed under the eyes of their deſigners, they were probably in better keeping. The Leaſowes, however, is now as well kept, perhaps, as it ought to be; and there is nothing ſtrikingly negligent at Hagley. They have both

of

of them reached that ſtate of maturity, when a poliſhed neatneſs is leſs required, than it is during the early bloom of embelliſhed places.

SECTION THE SIXTH.

ENVILLE.

FROM Hagley we proceeded to ENVILLE, the ſeat of the EARL OF STAMFORD, in the ſame picturable diſtrict; leaving with reluctance a lovely view of Shropſhire, as ſeen from the inn garden at Hagley; one of the moſt pleaſing views this diſtrict had afforded us.

ENVILLE, in ſituation, is ſimilar to Hagley and the Leaſowes. The immediate ſite is the precipitous face of an extended hill, broken into furrows, and watered by rills; of which there are two, as at the Leaſowes, that unite near the houſe, at the foot of the ſlope. The ſite of Enville is the ſteepeſt, moſt lofty, and largeſt of the three: containing ſeveral hundred acres, divided chiefly into ſheep walk and coppice wood, with kept grounds near the houſe, and with meadows and arable lands round the church and village, in the plain below.

In viewing these grounds, we were led to a summer-house-like building, at the immediate foot of the hill. It is situated upon the head of a small piece of water; beneath it, is a boat house; over it, a whimsical room; with a large painted glass window, towards the water. Finding nothing here to entertain, we signified a desire to proceed; but the guide (blockhead he for not amusing us better, or we for being in so great a hurry in so hot a day) informed us that a person had been sent to let off the cascade: a piece of information which, after what we had hitherto seen of cascades, was no great inducement for us to delay. Presently, however, the window was thrown open; and the most brilliant scene we had ever beheld presented itself. A SHENSTONIAN CASCADE, in full flow and fury; foaming and bellowing, as if the mountain were enraged: pouring down a river of water, white as snow, and apparently so copious, as to render our situation alarming; lest the house and its contents should be hurried away with the torrent. Had this scene broken upon the eye, abruptly and unawares, our sensations might have been excited as strongly as they were, on the first sight of the rocks of Persfield.

This house should contain something which would amuse every one, until the waters were laid on, The pool should be better covered from the walk, in in approaching it, and the lower part of the window be darkened, so that no water might be apprehended. If the opposite end of the room were first opened, to let in a view of the meadows, and tame country on that hand, it would not only help to amuse, but the contrast would assist in rendering the cascade scene the more striking.

THE splendour of the water is greatly heightened by the laurels and darker evergreens, which stretch out their branches from the rugged banks of the furrow, or shallow dingle, down which the water is precipitated: the foam, and the spray which flies from it, here mixing with the foliage of the evergreens, and there spreading over stoney surfaces; the steepness, the height, and the happy exposure of this fall; with the well judged distance at which it is placed from the eye; unite in rendering it one of the most sublime productions the hand of *Art* has effected.

ORIGINALLY, a chapel shewed itself at the top of this cascade, as the rotundo now does over that of Hagley. Fortunately, however, it is, at present, hid in wood; so that nothing but water, wood, and apparent rock, now enter into the composition of this fascinating scene. We could have looked on it long, with rapture, had not reflection brought to our mind, that the reservoir was emptying! This mischievous

miſchievous idea broke in upon our tranſports, and had nearly turned the whole into ridicule; until mounting the ſteep, examining the channel, and perceiving that, in ſome places, the water rolled over the dear native rock, a gleam of admiration returned.

THIS wonderful piece of machinery (for ſuch it may well be ſtyled) receives its rapid movements from one ſmall fountain; which alſo ſupplies a cold bath, recluſely ſituated above the reſervoir, which ſtores up its treaſures, for the liberal purpoſe of beſtowing them with greater profuſion on the ſtranger who may aſk ſo fair a boon.

CROSSING the head of the dingle, above the cold bath from whence the miracle-working water iſſues, the viewer is judiciouſly led to the edge of the wood, where ſome lovely views break abruptly upon him; compoſed of the Clent, Hagley, and Wichbury hills;—with the finely broken country about Stourbridge—uniting with the grounds of HIMLEY,—the reſidence of LORD VISCOUNT DUDLEY.

REENTERING the ſhade, we climbed a ſteep path, through an extenſive tract of coppice, until we reached the upper ſheep walk; a wide expanſe of naked turf; ſaving ſome tufts of hollies and

and a few ſcattered trees; containing ſome hundred acres, ſufficiently extenſive to maintain ſeveral hundred ſheep.

TOWARDS the center of this fine down, ſtands a white building,—the ſhepherd's lodge;—in which the ſhepherd and his family reſide. The principal part of it, however, is fitted up as a lounging room and obſervatory, for which it is ſingularly adapted. In elevation and expoſure, it reſembles Bardon hill, in Leiceſterſhire; which hill, it ſeems, is diſcernible from this place: from whence, and from different parts of the down, may be ſeen, on the other hand, the Wrekin and the Welſh mountains, with the Malvern hills, and the hills of Glouceſterſhire, &c.

THIS building, however, does not appear with full advantage. It is too large, and too conſpicuous, for a ſhepherd's hut; and too low and ill placed, as an obſervatory. A round tower, on a more elevated part of the down, would command no inconſiderable portion of the ſurface of this kingdom; and could not fail of being inſtructive, as well as entertaining, to thoſe who make geographical obſervation a part of their ſtudy, and one of their objects in travelling.

IT would be equally reaſonable, in the admirers of recluſe landſcape, to cavil at the practical botaniſt,

niſt, for being gratified and inſtructed by the diſtinguiſhing characters of a plant, as to cenſure the practical geographer,—one, whoſe favorite purſuit is to trace the greater outlines of the face of nature,—for being entertained and informed, on viewing the diſtinguiſhing features of his native country,

LEAVING the upper ſheep walk, we broke through a freſh part of the wood, into the further valley; a lovely well ſoiled glade; the fatting ſheep walk; which aſſimilates, in this point of view, with the grounds of HIMLEY; theſe ſiſter places happily playing off their charms to each other.

BELOW this, in a recluſe part of the coppice, is a ſmall ſequeſtered lawn, with a cottage and an aviary (apparently ill placed) with wild peafowls in the woods. And, below this, the lower ſheep walk, a plain incircled with wood.

WE now climbed the further ſide of the valley, to the upper ſhrubery; where we were more than recompenſed, by ſome ſtately Pines,—towering to the ſkies, and feathered to the graſs; and, from hence, a kept walk and a border of ſhrubs led us down to the lower ſhrubery: delightful ſpot! The Pines, here, are not only clothed to the graſs, but ſpread their

their mantles on the ground! and two ſiſter Limes are in full dreſs negligees, with trains flowing ſome yards from their conical outlines *: with a profuſion of beautiful ſhrubs, riſing out of the ſofteſt turf we ever ſaw: we had not conceived that graſs and trees, alone, were capable of producing ſo much richneſs and elegance. At the lower end of this ſhrubery, the houſe is ſituated.

WHAT a charming reſidence! No wonder Lord S. ſhould ſpend ſo large a portion of his time at Enville. But he gratifies not himſelf alone. His Lordſhip's liberality is equal to his taſte. His gratifications are heightened by thoſe, even of the mereſt ſtrangers, who ſeek enjoyment in his place: giving orders that nothing may be omitted, which can afford them gratification.

FROM what we could gather, on the ſpot, ENVILLE was originally deſigned by MR. SHENSTONE. The Caſcade and the Chapel are ſpoken of, with confidence, as his; but much has been done by others. MR. GREY, LORD STAMFORD's brother, has, of late years, done a great deal, and with good effect.

BUT

* This ſtriking appearance, perhaps, has been produced by the lower boughs that reſt upon the ground, having received from it additional nouriſhment.

But the high ſtate of preſervation, in which it is at preſent ſeen, and which ſets off the deſign to great advantage, is probably due to the attentions of Lord Stamford, himſelf; and to the aſſiduities of his preſent gardener; a man in years, and, we underſtand, of high reputation in his profeſſion; and who has probably executed much of what now appears with ſuch admirable effect.

In returning from Enville, we made our way by Himley; a place laid out on a very extenſive ſcale, by Brown; but we had only juſt time enough to ſee ſo much of it, as to determine us to take ſome other opportunity of examining it with due attention.

It is ſomewhat remarkable, that, within the compaſs of a few miles, there ſhould concenter four places of ſo much celebrity as Himley, Enville, Hagley, and the Leaſowes.

General Observations.

WHAT practical ideas have we collected in this little tour?

At the Leasowes we have learnt, that a few common paths, judiciouſly conducted, and a few ordi-

ordinary benches, judiciouſly placed, go a great way towards EMBELLISHING A FARM. Removing the more ſtriking deformities, diſcloſing hidden beauties, whether in the ſite or the offscape, and ſhewing them to the beſt advantage, in ſuitable walks, and reſting places, will generally make up the ſum of required embelliſhments; eſpecially in a place where much fortuitous wood abounds.

THERE, too, we ſaw the delightful effect of a ſimple path, leading through a RECLUSE DINGLE; and the abſurdity of attempting a CASCADE in a tame ſituation; and, generally, that the NATURE OF THE PLACE is ſacred.

AT HAGLEY we have ſeen the charming effect of a rich graſſy GLADE, deeply indented by the margin of a hanging wood; and that a SHEEP WALK, broken by maſſes, and diverſified by detached groups, is a ſuitable firſt diſtance to ſuch a view.

WE have alſo ſeen, in the ſame view, that an OBELISK may be ſo formed, and ſo ſituated, as to be ſufferable in Rural Ornament. We are of opinion, however, that the ſcene in which it appears ſhould, in ſome degree, be poliſhed, and that the ſky, alone, ſhould be its background. The idea of of ſimple nature, in a ſtate of neglect, muſt ever be done away, before poliſhed architecture can appear with good effect. And we are of opinion, that the obeliſk at Hagley pleaſes, in ſtanding forth boldly, yet modeſtly, and declaring, that the ſcene in which it appears, is not merely fortuitous, but is confeſſedly a work of taſte.

BUT the TEMPLE OF THESEUS, thruſting its proud portico into a ruſſet ſheep walk, and from out of a thicket of mean looking firs, on the contrary, diſpleaſes: not only as being out of place; but as holding out an oſtentatious diſplay of art, in a place where art was little wanted, and where it has been little uſed. Had this temple ſhewn its ſumptuous columns, in the face of the ſhrubery, which forms one confine of the beautiful glade, the foreground of this intereſting view, — in a part where taſte has done much, and where it ought to do its beſt, as being immediately under the windows of the houſe,—it would have appeared in place and character. What a charming effect a taſteful portico would produce, in the ſhrubery of ENVILLE! If Lord S.'s intentions are to pull down the preſent building, for the purpoſe of erecting ſuch a ſuitable accompaniment to his Spruces and Limes, we could forgive him.

IT

IT ſtrikes us, forcibly, that all buildings ſhould be in uniſon with the immediate ſite, in which they are ſeen: a principle, however, which does not appear to have been anywhere carried into practice; nor have we met with it, in theory: this TEMPLE OF THESEUS is praiſed by various writers.

AT ENVILLE, we have ſeen the grand effect of an ARTIFICIAL CASCADE, where the ſite is favorable, and where nature has furniſhed the groundwork. Much, however, of the faſcinating power of theſe ſplendid deceptions, may ariſe from their novelty, and were they common, they might no longer continue to pleaſe. But we are of opinion, that twenty ſuch as that of ENVILLE, ſcattered over the face of *this* kingdom, where natural falls are rare, would not pall the eye, nor *really* offend the feelings, even of men of the fineſt taſte; while, to men in general, they would be ſources of high delight.

AT ENVILLE, too, we have ſeen, that, by means of coppice wood and ſheep walk, a hilly broken country may be rendered highly ornamental, without exceſſive coſt. The coppices and ſheep walks of ENVILLE are ſaid to pay as much, now, as they did, when let off to farm tenants.

WILD PEAFOWLS are a beautiful accompaniment, in extenſive grounds.

BUT A COTTAGE, buried in extenſive woods, is out of place. Cottagers are ſocial beings. A hermit's cell, eſpecially if it were occupied, would be more in character.

AN AVIARY OF FOREIGN BIRDS appears to be equally ill placed, in ſuch a ſituation: exotic birds are apt accompaniments to exotic plants; and a ſhrubery, rather than a ſequeſtered dell, ſeems to be the moſt natural ſituation for an aviary.

IN the POLISHED GROUNDS of ENVILLE, we have ſeen what elegance and beauty may be produced, by trees and ſhrubs, judiciouſly diſpoſed, in grounds gracefully outlined, and on lawn highly kept.

UPON the whole, it is evident, from a view of theſe three places, that a SITE, naturally bold and picturable, may be rendered ornamental, at a ſmall expence, comparatively with that which is requiſite to the embelliſhment of a place, whoſe ground is tame, and whoſe features are inexpreſſive. How little has been done at ENVILLE! how much at FISHERWICK! and how much more at STOWE! BROWN's talent ſeems to have been peculiarly adapted

adapted to the embellishment of tame sites; giving a degree of character and expression to still life. SHENSTONE's forte, on the contrary, lay in setting off the stronger features of Nature, to advantage. It is possible, however, that education, rather than natural genius, led them into these separate walks. Be this as it may, BROWN's has been the most laborious, and, upon the whole, the most useful, part. A country, abounding with natural advantages, wants little assistance of art. But, where a large estate, and a principal residence, lie in a situation unfavored by Nature, or disfigured by fortuitous circumstances, an art which can create beauties, and hide or do away deformities, becomes highly valuable.

DIVISION THE FIFTH.

MINUTES

IN

PRACTICE.

TO the foregoing remarks on places, that have been improved by different Artifts, we add fome obfervations and reflections that have arifen out of our own experience, in places of different natural characters.

THESE places we were led to, in purfuing a PLAN FOR PROMOTING AGRICULTURE; which was firft brought forward, about fifteen years ago, and which has fince been extended to the MANAGEMENT OF WOODLANDS, and of LANDED PROPERTY in general: thus uniting, in the fame defign, the feveral branches of RURAL ECONOMY.

THE execution of this plan has been the leading object we have kept in view, fince the time it was firft first proposed; and we have, at length, the satisfaction to find, that the most difficult part of our labor is past. The SURVEY of the ESTABLISHED PRACTICES of ENGLAND has been made. Those of the *Eastern*, the *Northern*, the *Western*, and the *Central* parts of it, are before the Publick *. That of the *more Western* counties is now nearly ready for the Press, and the materials relating to that of the *Southern* counties, are collected, and will be prepared for publication, with all convenient dispatch.

SECTION THE FIRST.

MINUTES IN THE MIDLAND COUNTIES.

THE first of these places, in point of time, which engaged sufficient attention to give rise to written remarks, on RURAL ORNAMENT, was a small place in the MIDLAND COUNTIES †.

* See the list of Publications, at the close of these Volumes.

† See the Advertisement to the second Volume of the RURAL ECONOMY of the MIDLAND COUNTIES; also the Subject PLANTING, in the first Volume of that Work.

DESCRIPTION OF THE SITE.

THIS ſmall place is ſituated in a rich cultivated country, whoſe ſurface is ſufficiently billowy to admit of beauty; with an offscape, though not ſtriking, ſufficiently intereſting to accord with the gentleneſs of the ſite. The country is in a ſtate of incloſure, and much of it in high cultivation; with a few woods ſcattered thinly over it: a ſpecies of country which is very common, in the richer diſtricts of *this* kingdom.

SUCH MINUTES, made at this place, as relate more particularly to USEFUL PLANTATIONS, and the MANAGEMENT OF WOODLANDS, appear in the RURAL ECONOMY of the MIDLAND COUNTIES, publiſhed in 1790; ſuch as relate to RURAL ORNAMENT, and are conceived to be ſufficiently intereſting to bear the public eye, are inſerted here.

MINUTE THE FIRST.

1785. NOVEMBER 3. In ſtudying the nature of this place, with reſpect to RURAL ORNAMENT, ſome general ideas of practice have ariſen.

THE

THE MIDDLE GROUNDS and DISTANCES are the firſt ſubject of ſtudy. The beauties and deformities,—the pleaſing and unpleaſing objects,—which the more diſtant parts of the ſite and the ſurrounding country exhibit, or are capable of exhibiting, to the houſe, or other PRINCIPAL PLACE OF VIEW,—are the DATA. As in this caſe, for inſtance, ———, ———, ———, ———, &c. &c. are the objects on the middle grounds, and in the diſtances, which require to be expoſed to view; ———, ———, ———, &c. thoſe which ought to be ſcreened. Theſe are the UNALTERABLE DATA in the MIDDLE GROUNDS and DISTANCES.

BUT there may likewiſe be UNALTERABLE DATA within the NEARER GROUNDS, or more immediate environs of the houſe; ſuch as buildings which cannot with prudence be removed; or objects more deſirable than thoſe which they hide, and which, of courſe, ſhould be reſerved for INFERIOR POINTS of VIEW.

BEFORE any ſtep can be undertaken, with prudence, the ſeveral data, whether in the diſtances, or on the near grounds, ſhould be accurately aſcertained, and their relative ſituations, with reſpect to each other, be faithfully delineated; eſpecially if the near grounds be much wooded. For the ſame ſcreen may cover a defect, as well as a deſirable

 object;

object; a fact which it may be too late to have ascertained, when the screen is removed. A map of the place to be improved, with every knoll, water, and building, and with every mass, group, and single tree, accurately marked upon it, is essential to common prudence. If, from some elevated situation, the more distant objects can be seen, and lines be drawn upon the map, there spread out horizontally, to the several surrounding data, a degree of certainty will be obtained.

THE desired distances being let in, their respective near-grounds require to be moulded to them, as far as the given materials will allow, so as to throw them, when circumstances will permit, into PICTURABLE COMPOSITIONS; or, at least, into PLEASING VIEWS. Where offensive parts are to be hid, by fresh planting, much may be done, at the same time, towards uniting the near-grounds with the distances, so as to harmonize the compositions.

IF the near-grounds are naked of wood, and the views, in consequence, too broad,—masses of planting thrown in, so as to divide them into separate compositions,—each a PICTURABLE EYEFUL,—such as the eye can compass and repose upon with satisfaction,—may be productive of great improvement, at a small expence.

A BREAK

A BREAK of this kind, riſing at the angle of a houſe, aptly divides the views from the ſeparate fronts. If, at the ſame time, it can be made to hide a deformity, as it may in the inſtance under notice, and lay open two well featured diſtances, eſpecially if they happen to be "of various view", its operation will, of courſe, be ſtill more fortunate.

IF the angle of a houſe, requiring ſuch a break, happen to be the angle of approach, the planting ſhould be formed at ſuch a diſtance, as to admit of the road to paſs between it and the houſe; and to be contrived in ſuch a manner, as to ſcreen the buildings, until they break upon the eye, at once; thus effecting a two, three, or fourfold purpoſe. If, on the contrary, the angle of the building is not immediately approached, but is ſheathed in the kept grounds, as in this inſtance, the planting may be brought within a few paces of it; ſo as to bring the gayer tribes of ſhrubs, within a near view from the windows.

A BREAK of this intention ſhould ever accord, in diſpoſition and character, with the diſtance, the accompaniments of the near-ground, and with the character, and even furniture, of the room from which it is ſeen *; and, of courſe, ought to have ſide fronts as different as the views which they aſſiſt

* See the Section, PRINCIPAL RESIDENCE, page 285.

aſſiſt in forming. For, as far as general principles are admiſſible in works of taſte, the embelliſhments of the immediate environs ſhould be rendered ſubſervient to the houſe, the more diſtant parts of the ſite, and the ſurrounding country; ſo as to blend them, inaſmuch as they are capable of being blended, into one harmonious whole. And the immediate environs, being moſt under command, are the beſt inſtrument of union.

MINUTE THE SECOND.

NOVEMBER 5. In deſigning a ſcreen for —— it is a moot point, whether the plantation ſhould be formed againſt the fence, or be placed at ſome diſtance from it. In the firſt caſe, the expence of fencing will be leſſened, and the encumbrance to the ground will be leſs. But being thus fixed to the hedge, it will appear the ſame lifeleſs object from every point; whereas, in the latter, it will give variety, at each ſtep acroſs the view: and this holds good, in general. DETACHED MASSES OF WOOD, as well as GROUPS and SINGLE TREES, give a kind of animation to a ſcene: and this may be the reaſon why BROWN was ſo laviſh of them. But a *croud* of clumps, as a profuſion of ſingle trees, muſt ever disfigure the ſcene they appear in.

MINUTE THE THIRD.

NOVEMBER 6. The foot of the ſwell in —— is an obvious ſite for a BEND OF WATER. The ſkirts of the hill are naturally formed to give the water the appearance of winding, with eaſe, down the valley, towards another ſite, equally apt for another curve below; which two curves, as ſeen from the houſe, would have the effect of a natural river, eſpecially if they were judiciouſly backed with wood.

THIS appears to be the principle on which artificial-rivers ſhould be formed: not to expoſe lengthened canals; but to ſhew proper bends, at ſuitable diſtances, and in ſituations where ſuch *turns* might naturally be expected: after the manner in which we ever ſee a natural river with the beſt effect. Such as the winding eſtuary of the Severn exhibits, as ſeen from May Hill.

RIVER BENDS, in a rich view, are as diamonds in a rich dreſs. But artificial river bends, ſheathed in wood, are deceptions. They are ſo; the fraud, however, is of a venial nature. They are not ſo much intended to impoſe upon the viewer the idea of a natural river, as to give artificial waters their

full

full advantage; which cannot, in a rich cultivated ſite, be done in any other way, than in the character of a river.

THERE are numberleſs ſituations in this iſland, ſimilar to that under view; namely, a dip or ſhallow valley, with a rill falling down it; and the means of turning them to advantage are obvious, and not expenſive; as large extents of water are not required.

MINUTE THE FOURTH.

NOVEMBER 13. The RURAL ARTIST ſees trees in a different light to the BOTANIST. The ſhape of the leaves, the number of petals, and the parts of fructification, are to the artiſt, conſidered merely as ſuch, matters of ſmall importance; while, to the mere botaniſt, they are every thing worthy of his notice. On the contrary, the colour of their leaves in ſummer and autumn, and of their bark in winter; their times of foliation and diſleafing; their manners of ſhooting, the ſtructure and denſity of their heads, the outlines they uſually take, and the heights to which they aſpire, circumſtances little attended to by the botaniſt, are the properties moſt worthy of the attention of the artiſt.

MINUTE

Minute the Fifth.

November 18. A Screen Plantation, open on both ſides, ought not to be leſs than two rods (eleven yards) wide. Timber trees ſhould not be planted at leſs diſtance than half a rod, from the young hedge plants. Flowering ſhrubs, however, may be planted between them; to give beauty and fullneſs to the ſcreen, in the firſt years of its growth.

Minute the Sixth.

November 18. The effect of a plantation, diſtant from the point of view, cannot be accurately judged of, before the ground be broken, or ſome other obvious diſtinction of colour take place.

In lining out the projection and receſs,—the *promontory* and *bay*,—of plantation E. the outline determined upon appeared, on the ſpot, to be perfectly ſatisfactory: but now that the ground is dug, the planting begun, and the lines rendered diſtinct from the houſe, the point of the promontory is evidently too ſharp.

In

IN a more diſtant, more extenſive, and more fortuitous ſcene, where the picturesque rather than the beautiful is required, a ſharpneſs and even raggedneſs of promontory may be in character; but, here, art is obvious, and gracefulneſs of outline is required.

IN lining out plantations at ſome diſtance from the point of view, as four or five hundred yards, ſome conſpicuous mark is requiſite, and nothing is preferable, perhaps, to white or light coloured hurdles; which are conſpicuous and readily moved.

MINUTE THE SEVENTH.

1786, JANUARY 25. In LINING OUT NEAR-GROUNDS, the firſt ſtep is to note the data;—to mark the GIVEN POINTS. In joining theſe points, place the intermediate marks at equal diſtances; or the eye will be deceived. If the line be long, trace it repeatedly with the eye, from each extremity; and having, by theſe repeated tracings, rendered it familiar; and having as repeatedly trod it out, in contrary directions; let an aſſiſtant follow with even ſtrides, while a third perſon place marks at every ſecond, third, or fourth ſtep, according to the length and flexure of the line.

ON

ON broken ground, or while ſnow lies on graſs-land, the footſteps of the deſigner are ſufficiently obvious, as a guide to the marker; but on green turf, it is requiſite to fix them, in the inſtant, by permanent marks.

THE line being thus made conſpicuous, it requires to be examined, from every point of view, and every walk and pathway, which commands it; and if it conſiſt of more than one part or diviſion, occaſioned by different given points, each part ſhould be made to play into the other, ſo as to render it agreeable to the eye, from whatever point it may be ſeen.

IN highly poliſhed grounds, immediately under the windows of an elegant room, the ſmalleſt deviation from the line of beauty offends the eye *.

MINUTE THE EIGHTH.

JANUARY 26. In forming mixt ORNAMENTAL PLANTATIONS, ſome plan of proceeding is requiſite

* For further remarks on this topic, ſee — "A REVIEW OF THE LANDSCAPE, a didactic poem; alſo of An ESSAY ON THE PICTURESQUE; together with PRACTICAL REMARKS ON RURAL ORNAMENT,"—page 221.

ſite to be laid, reſpecting the plants, previouſly to the commencement of the operation.

THE ſpecies of plants being determined upon, and the requiſite number aſcertained, it is proper to liſt them, agreeably to their reſpective heights of growth, in this climate * ; in order that the talleſt growers may be placed in the rear ranks, the lower towards the front.

TO aſſiſt in the due ARRANGEMENT, whether as to *height* or *colour*, collecting twigs or ſmall boughs from the ſeveral plants (that is, as many ſlips as there are plants of each ſpecies), and diſpoſing them agreeably to the intentions of the artiſt, previouſly to any of the plants being put in, will be found beneficial; as ſaving much ſuperintendance and labour, and preventing the plants themſelves from injury, in being dragged about, from place to place, before their proper ſituations are found.

MINUTE THE NINTH.

JANUARY 29. In DESIGNING, whether in compoſition or in detail, anxiety and exceſſive poring over

* For which purpoſe a liſt of trees and ſhrubs, arranged agreeably to their growths, in this country, will appear in the ſecond Volume.

over the ſame ſubject, ſerve only to vex and fatigue the imagination; rendering that irkſome, which ought ever to be pleaſurable. It will generally be found, perhaps, that ſauntering over the field of improvement, and bending the mind to ſuch ſubjects as riſe ſpontaneouſly, will be more productive of practical ideas, adapted to the nature of the given place, than any preconcerted plan of ſtudy. Even in the detail, returning repeatedly to the dubious point, with the mind unbent, will frequently unravel the knot, and clear up the doubt, ſooner, than intenſe unremitted application.

MINUTE THE TENTH.

JANUARY 31. It not unfrequently happens, that a pleaſing object, and one which is unſightly, appear in the ſame line of view from a principal point; as —— and ——

IF the deſirable object appear in the offscape, and much above the eyeſore, as in this caſe, the evil is to be remedied, by hiding the offenſive part with ſhrubs, of a natural growth, ſufficiently high, to operate as a ſkreen to the deformity; yet ſo low, when at their fulleſt height, as not to hide the diſtant object.

SHOULD the diſagreeable object be in the off-ſcape, and the deſirable one upon the nearer ground, tall-ſtemmed trees would hide the one, without ſhutting out the other, entirely from the view.

MINUTE THE ELEVENTH.

FEBRUARY 2. In forming A SIDE SKREEN, where a line of tall-grown trees are the given back ground, or rear rank, ſome cautions are required.

IT is particularly requiſite, in this caſe, to liſt the moveable plants that can be commanded for the purpoſe; not only, according to their natural growths, but their actual heights, at the time of planting; for, if ſome ſhow of proportion is not preſerved, ſo as to bring down a ſlope from the tops of the growing trees to the gravel or turf which ſhall embrace the foot of the ſkreen, ſufficiently regular not to offend the eye, the deſign muſt be marred in the firſt ſtage of execution. For, with all the precautions which art can furniſh, a plantation of this deſcription muſt remain unſightly, for a few years after it is formed. The tranſplanted trees require to be thinned and lightened of of their boughs, ſo as to proportion them to the length and number of tranſplanted roots, or their ſucceſs will be uncertain. And, with every precaution, their progreſs, for a few years, until they have eſtabliſhed themſelves in their new ſituation, muſt neceſſarily be ſlow. Hence, at the time of planting, their tops, as ſeen from the principal place of view, ſhould not only appear thin and unfurniſhed, but ſhould riſe above the general line of aſcent; in order to allow for the ſuperior upward progreſs of the eſtabliſhed plants, during what may be termed the naturalization of the ſtrangers.

To guard againſt the incurſions of the eſtabliſhed trees, as well as to check their upward growth, their roots, on the ſide next the planting, ſhould be cut off at a ſuitable diſtance from their ſtems, at the time of double digging the ground to receive the freſh plants: and moreover ſhould, from time to time afterward, be prevented from injuring their weaker neighbours, by over-running the paſturage of their yet feeble roots; which ought, for ſome years, to be defended, likewiſe, from weeds and other enemies.

MINUTE THE TWELFTH.

FEBRUARY 2. IN TRANSPLANTING young trees, of eighteen or twenty feet in height, it is imprudent to attempt to take up more mold with their roots, than will with certainty adhere to them, until they are fixed in their new ſituation. For that which falls off in carriage, ſeldom fails of carrying with it ſome of the finer more valuable fibres; eſpecially if the ſoil be in any degree tenacious. Long roots, well furniſhed with fibres, and duly bedded in fertile mold, are better pledges of ſucceſs than heavy balls of ſtale earth; which, by rendering the plants cumbrous, and difficult to be moved, too frequently cauſes them to be bruiſed and maimed, in the operations of removal.

THE ſuperfluous mold ſhould be diſengaged (with the hands or a fork with round tines), before the plant be attempted to be lifted out of its place of growth: and, from this time, until it be placed in its new ſituation, the roots ought not to be touched with the hands.

PLANTS of this ſize are beſt removed, by means of a lever or long pole, guarded in the middle with ropes

ropes of hay or ſtraw, to preſerve the bark of the ſtem from injury. This guarded part of the pole being placed againſt the foot of the ſtem, the plant is pulled down upon it: two men bear up the root with the pole, while a third ſteadies the top, and keeps the plant horizontal, until it arrive at its place of deſtination; where, ſuffering the top to riſe, it willingly regains its erect poſture.

IN a dry ſeaſon, it is eſſential to common prudence, TO WATER THE PITS BEFORE THE PLANTS BE SET IN THEM; firſt returning ſo much of the beſt of the ſoil as may be judged neceſſary to ſet the plants upon. If, on examining the bottom of a given plant, when it arrives at the pit in which it is intended to be planted, too much or too little mold has been returned, or if the ſurface of the mold is not anſwerable to the form of the under ſide of the root, a perſon, attending for the purpoſe, ſhould make the neceſſary regulation, while the plant remains ſuſpended on the arms of the bearers; for it is ever miſchievous to a plant, to place and replace it, in the operation of tranſplanting; and ſuch unworkmanlike conduct is ever diſgraceful to a planter.

THE precautions neceſſary to be had in planting, are, to unite the freſh mold with the ſoil which has been removed with the roots, ſo as to form them into one uniform maſs, without any pores, vacancies, or interſpaces, between them; and, in effecting this, to bed the roots, and particularly the ſmaller fibres, evenly among pulverized fertile ſoil; leading them out, horizontally, or ſomewhat dipping, from the part of the nucleus or bulb of the root, from which they naturally iſſue; being mindful not to raiſe the mold too high before they are laid down, nor to force them down, before the mold is high enough to receive them; ſpreading them out wide, like fronds of fern, and tire above tire; endeavouring to diſtribute them equally among the mold; in order to give them equal ſpaces, or range of paſturage; but, in endeavouring to do this, not to cramp them, or wreſt them forcibly from their natural direction. If a root be longer than the reſt, and too long for the width of the pit, a notch ſhould be cut in the ſide of it, to give room for the root to lie eaſy, and at its full length, not more to aſſiſt in giving ſtability and firmneſs to the plant, than to enlarge the field of paſturage of its roots, in the firſt inſtance;—in the hour of need.

THE lower tire of fibres being bedded, in this manner, and covered over fully with mold (the thickneſs of covering being regulated by the ſituation of the next tire of roots) they ſhould be preſſed down firmly; firſt with the hands, and afterwards with the feet, to prevent any hollowneſs

or

or falſe filling, and, in conſequence, a ſettling of the mold; which would cramp the upper tires of the roots therein to be laid; and, at the ſame time, to give the greateſt firmneſs to the plants, at a time when much may depend on the undiſturbed ſtate of the fibres.

A well rooted plant, put in with due precautions; ſuch as packing in the freſh mold, by hand, while the plant is in a ſomewhat heeling poſture, ſo as to give freedom to the workman, and additional firmneſs to the filling; bedding the rootlets ſingly, firmly, and divaricated, among the ſoil (fine mold being ſcattered over the hands of the planter, while he keeps each branch in its proper place); treading layer after layer, as the pit is filled in; and, finally, loading the roots with ſoil;—receives an immediate firmneſs and ſtability, which, in ſheltered ſituations, precludes the neceſſity of ſupporters, even to plants of fifteen to twenty feet high: indeed, well rooted plants, thus put in, ſeem to ſtand firmer—ſtiffer—after planting, than before they were taken up.

WITH reſpect to the *pruning of the tops*, part of it ſhould be done previouſly to the removal; the finiſhing part being done after planting. Lightening the heads before tranſplanting (and eſpecially ſhortening the lower boughs of the Pine tribe), renders the plants better to handle, and ſecures them from ordinary winds, preſently after removal. But there is a twofold reaſon for completing the operation after the plants are ſet in their new ſituation. The additional top, probably, encreaſes the acting power of the fibrils, to feed in their new paſture; and, when the ſeveral plants are in their places, the deſirable form of the top of each, ſo as to make it aſſimilate with its neighbours, and give the beſt ſurface which a freſh plantation of this kind is capable of admitting, may be beſt ſeen.

It may be ſaid, in general terms, that the top of a plant ſhould not be touched with the pruning knife, while it is out of the ground; ſaving ſuch part of it as is out of the reach of the pruner, when ſtanding. The principal part of the pruning, whether of trees or ſhrubs, ſhould be done before the plants are taken up; the finiſhing given after they are replanted, and have begun to work in their new ſituation. But the leaders of tall plants ſhould be particularly attended to, while they are in a horizontal poſition.

The expence of tranſplanting is conſiderable. Three men moving plants, near twenty feet high, and as thick as the leg, in the above-deſcribed deliberate manner, and carrying them a hundred yards, do not move more than ſix or eight plants a day.

a day. This (with the previous expence of digging the holes), is not leſs than eightpence or ninepence a plant. It is true, by hurrying over the work, in a ſlovenly way, ſomething might be ſaved. But the ſaving, compared with the riſk of loſing plants of this ſize, the loſs of labour, and the disfigurement of a plantation of this kind, is no object of conſideration.

FOR further Remarks on this Method of Planting, ſee the RURAL ECONOMY OF THE MIDLAND COUNTIES, Minutes 146 and 168.

MINUTE THE THIRTEENTH.

FEBRUARY 6. ON TRANSPLANTING the PINE and FIR TRIBES, into *plantations*, or extended maſſes of wood, the points or leading ſhoots of their lower boughs ſhould be taken off. *Firſt*, to check theſe boughs, and thereby enable the roots to ſend up a better ſupply of nouriſhment to the leaders and upper boughs in general. For this purpoſe, if the lower boughs be numerous, they may be ſhortened, even to the innermoſt wings or pair of branches, with advantage: the Spruce Firs of plantation A. ſucceeded perfectly, the laſt ſeaſon, under

under this treatment *. *Secondly*, to prevent their encumbering their neighbours; the treatment being ſingularly applicable to the Scotch Pine, in mixed plantations. In almoſt every place, the evil conſequences of not attending to this are obvious. And *Laſtly*, in the interior of a plantation, the ſooner the lower boughs die and drop off, the more valuable the timber becomes.

BUT of the Pine tribe, ſet out as *ſtandards*, or in *groups*, or in the *outer ranks* of a plantation, the lower boughs are their beſt ornament. How rich is their effect at BERKLEY, at ENVILLE, and at FISHERWICK. But, even in this caſe, it is not always neceſſary, or proper, to ſuffer *all* the lower boughs to remain at their full length. If they are numerous, they will not only carry off too much ſap, and thereby weaken the head of the plant, but themſelves become ſlender, feeble, and take a buſhy unſightly form;—whereas, by leaving a proper number of ſtrong boughs, in ſuitable directions, and checking the reſt, the plant will at once be invigorated, and acquire variety of outline and ſtrength of feature, as it grows up.

TRANSPLANTED Roots can only ſend up a certain ſupply of ſap, and it is the planter's duty to

* *See* RUR. ECON. of the MID. COUNT. Vol. ii. p. 351.

to see that no portion of it be spent in vain,—that every drop be applied to the most useful purpose. And, further, inasmuch as single trees require a greater quantity of boughs to be left standing, the planter, if he even hope for success, ought to be sedulously attentive to take up, and remove with them, a quantity of fibres proportioned to the necessary exhaustion; and the greatest possible length of root, to give them stability and firmness, in their new situation.

MINUTE THE FOURTEENTH.

FEBRUARY 20. (see MIN. 6.) In an attempt to COLOUR this part of the plantation,—so as by rendering the recess dark, to throw it into shadow, and by giving a degree of lustre to the projection, give variety at least, if not picturable effect,—we perceive that the art of COLOURING WITH TREES is attended with a difficulty which we were not aware of: their winter and summer colours are not only different, but, in some valuable species, opposite. Thus the *Lime*, in winter, is remarkably dark, but, in summer, its leaves are of the lighter shade of green; and the *Esculus*, which is singularly dark, in summer, has now, a somewhat pallid appearance.

How-

However, there are other ſpecies, we find, which are well adapted to painting. The *Larch*, for inſtance, is ſingularly light, in winter; and, in ſummer, it wears a lively green. Again, the *Aſh* is uniformly light and elegant: the *Planes* and the *Aria* are ſtill more ſplendid, in both ſeaſons. But the *Evergreens* are the moſt permanent; though not altogether ſo; as, at the time of making their ſhoots, they wear a lighter garb, than at other ſeaſons. In winter, the Scotch Fir, and the Larch, are admirably adapted to colouring; and, in beguiling the dreary reign of winter, the ſkill of the artiſt is beſt employed. Hence, the back of the receſs is already a maſs of Firs, and dark deciduous trees; the projecting point to be made as ſplendid as Larches, Planes, and Arias can render it: meaning to aſſimilate and ſoften them off, by degrees, with the Beech, as a ſemi-tint or intermediate colour, to the Oak and the Eſculus.

But after all, painting with living colours, and in open daylight, is not only difficult, but in a degree unprofitable; for a beam of the Sun may turn the whole into ridicule; by throwing the light into ſhadow, and rendering the ſhadow a maſs of light.

In plantations diſtant from the eye, all colouring is improper; and in thoſe at hand, a fortuitous aſſem-

aſſemblage is, perhaps, on the whole, preferable to any ſtudied arrangement.

NEVERTHELESS, in ornamental plantations, in which plants of different heights are uſed, regard muſt be had to that circumſtance; and, in the more gaudy exotic ſhrubery, colour ought not to be wholly neglected. In winter, Evergreens mixed with the crimſon branches of the American Cornus, and relieved with the ſplendid foliage of the ſilvered tribe of ſhrubs, have a pleaſing effect.

THE ARRANGING OF PLANTS, however, whether as to colour or height, is a moſt tormenting employment. A Painter has his pallet and bruſh in hand, and his colours in paſſive obedience to his will. He ſees his picture at one view, or can run his eye over it, with a ſingle glance, and can, in a moment, make or unmake whatever his imagination dictates, or his judgment condemns.

BUT not ſo the Rural Artiſt; his colours are too unwieldy, to be worked up with his own hands: he is, of courſe, liable to the miſconceptions and aukwardneſſes of workmen; and he cannot correct an error without injury to his work. Beſide, his canvas is not ſet up before him, ſo that he can ſee the whole at once; nor can he ſketch out his whole deſign, in a few hours, or perhaps a few days: planting

planting is a progreſſive buſineſs, and is liable to ſeaſons and the weather; eſpecially if the ſite be of conſiderable extent.

FOR ſmall plots, aſcertaining and liſting the plants, and diſtributing boughs, in the manner already mentioned, is perhaps the moſt eligible. And, for larger plantations, dividing them into compartments, and proceeding in a ſimilar way, is the moſt practicable method we have yet been able to hit upon. Thus, the number and ſpecies of plants for the whole plantation being aſcertained; the number of each ſpecies, requiſite for each ſeparate compartment, muſt be found, and their boughs be diſtributed.

THE diſtribution of the marks is beſt done, before the holes are dug, where circumſtances will admit of it; as each ſpecies of plants may then have ſpaces aſſigned them, ſuitable to their reſpective natures and manners of growth; and the ſize of the pits, too, may be adapted to the probable length of root which each ſort is known to riſe with; the workman deſcribing a circle round the marking twig, and returning it to the center of the hole, when it is formed.

BY calculations of this kind, and by methods of this ſort, ſtrictly adhered to, moſt of the embarraſſments

ments incident to forming mixed ornamental plantations, may be avoided, much labour be ſaved, many plants be preſerved from injury, and the execution be rendered conformably to the deſign.

MINUTE THE FIFTEENTH.

MARCH 24. In TRANSPLANTING large plants, the ſucceſs depends, chiefly, on taking them up with a good LENGTH OF ROOT; which ought not, in ordinary caſes, to be leſs than one fourth of the height of the plant. It may be difficult, in moſt caſes, to take up twenty feet plants, with roots five feet long; but, where plants ſtand tolerably free, there is none in taking up plants of twelve feet high, with roots three feet long.

ROOTS are the natural and beſt ſtay of a plant; and a planter had better beſtow ten minutes in taking up, than five in ſtaking. It is not neceſſary that balls of earth, of a ſemidiameter equal to the length of the roots, ſhould be moved. Theſe may be reduced to any ſize. Indeed, the more experience we acquire in tranſplanting, the more anxious we become for roots, and the leſs ſo for balls of earth.

Theſe,

Theſe, however, are deſirable when they can be moved without exceſſive expence of carriage, and without injury to the roots.

MINUTE THE SIXTEENTH.

MARCH 30. A view may ſometimes be improved, at an eaſy expence. A few remaining trees, of one line of an avenue, had a bad effect, from the windows of a principal room, to which they nearly pointed, but not directly, their ſtems being ſeen diſtinct; and, of courſe, produced the bad effect of a ſtraight line of trees.

THIS defect was remedied by a ſingle ſhrub —a well furniſhed plant—about ten feet high, which covers the ſtems, while the tops take the form of a group; the idea of a line being loſt, in the general effect. How often may ſimilar defects be hid in this way. Had the width of the deformity been greater, a group, or a tuft of ſhrubs, would have been required.

MINUTE THE SEVENTEENTH.

APRIL 1. When ſhrubs have been drawn up tall, and rendered naked at the bottom, by being crouded

crouded in a nurſery, or a crouded plantation, it is almoſt impoſſible to prune them, into forms which will pleaſe the eye. A low growing plant, which has been drawn up tall, and conſiſts only of a few ſprawling boughs, ſpreading out like a fan, has been improved into a well looking ſhrub, by planting a low ſpreading ſucker, in the ſame pit, and placing it in front, and ſo as to fill up the central vacancy: the two affording, in this combined form, a well furniſhed plant: a venial fraud, which may frequently be practiſed with advantage.

MINUTE THE EIGHTEENTH.

APRIL 1. In PRUNING SHRUBS, at the time of tranſplanting, much may be done towards the future appearance, as well as the future ſucceſs of the plant. This is not to be effected by lopping off the ends of the twigs, in general, and thus giving the ſhrub the form of a cabbage; but by taking out the inferior branches, cloſe to the ſtem or the thicker boughs; and even taking out ſome of theſe, ſo as to make breaks in the outline;—will often give additional feature and elegance to the plant; while, by thus reducing the top, the roots are rendered the better able to ſend up a ſupply of ſuſtenance, to the parts which are left ſtanding.

MINUTE THE NINETEENTH.

APRIL 2. In tranſplanting ſhrubs which throw up SUCKERS, theſe ſhould be carefully laid aſide, and placed in a nurſery quarter, to acquire roots, and become a ſupply of plants, in future, at a ſmall coſt. Alſo, from neglected ſhrubs, which afford natural LAYERS, wherever the boughs touch the ground, each rooted twig ſhould be ſeduloully collected.

MINUTE THE TWENTIETH.

APRIL 2. In TRANSPLANTING top-heavy Evergreens, as Virginia Cedars, Junipers, Arbor-vitæs, &c. for STANDARDS, it is prudent to PLANT A SUPPORT with each of them. Not an ordinary ſtake, but a larger and more clubbed truncheon; placing the large end downward in the bottom of the pit, a ſtraight part riſing ſome few feet above the ſurface, and nearly cloſe to the ſtem of the plant; which being faſtened to it, by means of ſoft bandages, gains a ſeaſonable firmneſs, without any outward appearance of ſupport.

MINUTE THE TWENTYFIRST.

APRIL 7. IN LINING OUT WALKS, a ſlight covering of ſnow is advantageous, in ſhewing the track of the deſigner; which may be improved, as occaſion may require. Stakes, though proper in lining a plantation, as ſhewing at once the effect of the intended fence, or of the marginal ſhrubs, may tend to deceive the eye, in the effect of a walk; whereas a track, whether in ſnow, or on the ſurface of broken ground, or given by a ſharp inſtrument, drawn by a ſecond perſon, ſo as to ripple the ſurface of green turf, is in effect the walk; differing only in width, from the real walk when finiſhed.

IN wild or fortuitous ſcenery, the firſt devious tract will generally have the beſt effect. But, in highly embelliſhed grounds, it requires to be lined out, with ſcrupulous attention to the beauty and gracefulneſs, which ought to mark every line, in poliſhed ſcenery.

WHEN a walk winds acroſs a lawn, broken by tufts and relieves of ſhrubs, it ſhould appear as if attracted by the various beauties of the ſcene: it ſhould make boldly towards them, hang to their margins, and ſeem to leave them with reluctance.

IN tracing paths, through plantations of tall growing trees, intended to riſe into groves, the trees themſelves ſhould ſeem to direct the path, which of courſe ought not to be determined on, before the trees are planted. In plantations formed of tall tranſplanted trees, ſuch paths may be formed immediately after the trees are planted; otherwiſe, they ſhould be deferred until the trees are grown up, and the obſtructing plants be removed, in the thinnings: the direction of the path being *determined* (but not formally *marked)*, by evergreen underwood, as Holly, Privet, Box, or cuttings of Laurel; and a narrow pathway, no matter how intricate, may wind in among the young plants, for the purpoſe of rendering the plantation itſelf commodious, in viewing, thinning or pruning the plants. A path three feet wide is ſufficient for this purpoſe.

NARROW paths of this kind render a plantation commodious, and are formed at a trifling expence. The middle of the path is the natural ſurface of the ground, a ſloping channel being ſtruck with a ſpade on either ſide: this, and pruning off the boughs which ſhoot towards the path, affords the required accommodation.

MINUTE

MINUTE THE TWENTYSECOND.

APRIL 10. FENCES IN ORNAMENTED SCENERY. For the ſecurity of highly kept grounds, the FOSS, accompanied with maſſes and tufts of wood, is the moſt eligible; as giving the eye the leaſt reſtraint, and as ſerving beſt to aſſimilate the immediate environs of the houſe, with the contiguous park or paſture grounds *.

BUT, in the lower ſtyles of ornament, a leſs expenſive boundary is preferable: and for the fence of a plantation, not included within the limits of the kept grounds, but ſtill within diſtinct view from the houſe and its environs, a ſimple guard, ſufficient againſt paſturing animals, without being offenſive to the eye, is the only requiſite.

THAT which, after much conſideration, we adopted and executed, here, is a ſloping ditch and reclining bank, with a dwarf rail fence, *hanging* in the face of it, at ſuch a diſtance as to prevent cattle from climbing over it, and ſheep from creeping beneath it; and with a line of hedgewood on the inner ſide, when its uſe is to guard a plantation.

 THE

* See the REVIEW of the LANDSCAPE, &c. p. 231.

THE face of this fence may either be turned towards the plantation, or from it. In the former case, it is less visible; but in the latter, it is a firmer better fence, and incurs a less waste of land; for the slope of the foss being made easy, and sown with grass seeds, as well as the face of the bank, which also falls gently back, the pasturable surface is greater with this, than perhaps with any other fence. By adding a dwarf paling, this fence becomes effectual against hares, at a moderate expence.

As a fence against cattle and sheep, the following have been the dimensions, and manner of construction, here. Level the ground, and turn a gauge turf; dressing it with an even firm angle, as a guide to the whole work. Behind this turf, lay in morticed posts, four feet and a half to five feet long, placing the lower end of the mortice about eighteen inches from the angle of the gauge turf, and in such a position, as to form with the face of the bank, when finished, the lower point of an equilateral triangle, whose upper side is horizontal. Fasten the posts, and carry up the bank, with the excavated mold of the ditch; forming the face of the bank with turf; and ramming in the soil firmly behind it, as the bank is carried up; to prevent its settling too flat: and the more effectually to prevent this, the upper part of the bank should be built,

built, ſomewhat more upright, than the foot of it; which ought, of courſe, to form, with a vertical line, an angle of 30°. In this manner, the face of the bank is raiſed to about ſix feet ſlope, allowing ſome inches for ſettling; the length of ſlope, when ſettled, being about five feet and a half; namely, three feet below the level of the ground, and two feet and a half above.

THE rails are *ſlipt* in; the preceding one being bound by that which ſucceeds it: care being had not to jar the poſts, before the bank be firmly ſettled.

WHEN the plantation is up, or the hedge becomes a fence, the bank may be thrown down: the temporary fence having then done its duty.

NEARER the eye, and where a hedge would be unſightly, the rails and poſts may be repaired, from time to time, at an expence extremely trifling, compared with that of a wall or paling.

SECTION THE SECOND.

MINUTES IN DEVONSHIRE.

THE next instance of practice, in the Rural Art, occurred in Devonshire; at Buckland Place, formerly *Buckland Priory*; the residence of the family of Drake, from the time of the Circumnavigator, who purchased it, until the death of the late valuable possessor, Sir Francis Drake;—now a seat of Lord Heathfield.

Description of the District and Site.

The Western District of Devonshire, in which this place is situated, abounds with picturable scenery. It forms a sort of vale between the Dartmore and Cornish mountains; but differs from ordinary vale districts, in the abruptness of its surface and the dryness of its soil; possessing, in these respects, the distinguishing characters of an upland country; broken, in a striking manner, into ridges and vallies; and, in some places, rising in detached hillocks;

hillocks; thus giving infinite variety of *ground*, whoſe ſteeper hangs are moſtly clothed with wood; which frequently mantles down to the margins of the rivers and eſtuaries, with which the diſtrict is happily interſected.

BUT the immediate ſite under deſcription, though ſurrounded with ſcenery of the laſt mentioned caſt, does not partake of it. The houſe, ſituated in a dip or ſhallow valley, is beſet with well turned knolls, folding with each other, in a beautiful manner. The whole is well ſoiled, and in a ſtate of cultivation, except the more diſtant ſwells, which are ſteeper than the reſt and hung with wood; over which appears a riſing knoll of heath, forming a happy offscape, to the principal view from the houſe: altogether, a *monaſtic* ſite.

SOME ſixty years ſince, much grove planting had been done about the houſe: and, during the laſt twenty or thirty years, the whole had been ſuffered to grow up in a ſtate of neglect; ſo that the houſe might be ſaid to ſtand in a valley of wood, and to be rendered, at once, unpleaſant and unwholeſome.

SOME alterations, however, had taken place, twelve or fifteen years ago, cloſe about the houſe, within the walls of the old garden: the terraces having

having been thrown down, and the ground formed and laid out, agreeably to the modern ſtyle of ornament; and in a manner which would have done the artiſt credit, had the houſe been modern; but, to the remains of the old Priory, terraces and graſs plots were the beſt accompaniments.

PRELIMINARY REMARKS.

WITH theſe data, there was only one line of procedure. The character of the foreground, as well as of the diſtances, was beauty; and all that art could do, with effect, was to bring the middle grounds into uniſon with them: to break the groves and ſkreens, in ſuch manner, as to leave well formed maſſes of wood, with viſtas and graſſy glades between them; ſhewing, with the beſt effect, the beautiful undulations of ground, with which the ſite abounds; but which were almoſt wholly ſhut out from the houſe.

THIS has in part been done; not more with the view of diſcloſing the beauties of the place, than to ventilate it, and thereby endeavour to counteract the exceſſive moiſtneſs of its climature: an extent of orchard ground, ſpreading over the valley below the houſe, with ſome fences which

disfigure

disfigure one of the boldeſt ſwells, ſtill mar the principal views; and while the preſent *purpoſe* of the place remains, it might be wrong to remove them: ornament and utility ſhould ever go hand in hand, and be ready to accommodate and aſſiſt each other, as the exiſting circumſtances of the place require.

ON the ſubject of RURAL ORNAMENT, we find very few MINUTES at this place: our attention being principally bent towards the ſubjects of RURAL ECONOMY; particularly towards the improvement of the noble, we had almoſt ſaid magnificent FARM, which, at preſent, may be ſaid to conſtitute the place itſelf*. Indeed, the ornamental improvements being chiefly confined to the dividing of continuous ſcreens, and an alteration in the line of approach, the ſubjects of memorandum were few.

NEVERTHELESS, the difficulties of opening picturable † viſtas, through lines of tall grown wood, are ever too great, and their impreſſions too ſtrong on the mind, to paſs away wholly without notice.

MINUTE

* Some account of this Farm will appear in the RURAL ECONOMY OF THE WEST OF ENGLAND; now nearly ready for the preſs.

† PICTURABLE,—grateful to the eye, in nature, and capable of being repreſented, with good effect, in a picture.

MINUTE THE TWENTYTHIRD.

THE practical ideas that grew out of the experience which this place afforded, turn chiefly on the proper ſeaſon for this operation; and on the extreme caution and continued ſtudy requiſite to the due performance of it. Endeavouring, before the work is ſet about, to gain a general idea of the effect of each opening, from every point of view which will command it, is the groundwork of ſucceſs. This may be done, at any ſeaſon; but there are only two, in which the operation itſelf can be proſecuted with full advantage. This is either in autumn, while the leaves are changing their colours, or in the ſpring, during the progreſs of foliation. The latter is the moſt proper ſeaſon: for the ſtructure, as well as the outline, of each tree may then be diſtinctly ſeen. During ſome days, accordingly as the progreſs of vegetation is ſlow or rapid, ſcarcely any two trees, even of the ſame ſpecies, are exactly of the ſame colour: while one retains its wintry hue, another is forming coloured buds, a third is in fuller bud, a fourth burſting, a fifth in pallid leaf, a ſixth of a deeper tint, &c. &c. ſo that, at this critical juncture, the branches of adjoining trees may, in general, be ſeen

ſeen diſtinctly, how intimately ſoever they may be mixed with each other: of courſe, the outline of either may be ſeen, before the other be removed.

MINUTE THE TWENTYFOURTH.

OTHER ideas, which it was thought right to memorize, relate to TUFTS OF HEDGEWOOD, left ſtanding on cutting tall-grown Hedges; to break the meagre monotonous lines of farm fences; in which, as is common in this diſtrict, no hedgerow timber appears.

THE ſapling ſhoots from the ſtools, or old roots, of the *Aſh* run up tall, and take better outlines than thoſe of the *Oak*; which, on the high hedge mounds of Devonſhire, generally ſpread too wide, and take an outline too rotund and ſquat: the ſapling groups of the *Cheſnut*, the *Wild Cherry*, and the *Mountain Sorb*, alſo take deſirable outlines; eſpecially when Blackthorn, or other ſlow-growing ſhrubs, happen to ſtand on their margins. With a little attention to the freeing of the baſes of theſe ſapling tufts, while riſing, their forms would be more natural, and their effect more pleaſing.

THE

THE forms and ſizes of theſe hedge tufts ſhould be as various as the circumſtances which give riſe to them. Tufts, of every dimenſion, feathering to the bank; taller groups, riſing with naked ſtems; and even ſingle ſtems, if ſuitably furniſhed; may afford variety and richneſs to the ſcenery, as well as that ſhade and ſhelter, which a newly fallen hedge, hacked down from end to end, is rendered incapable of furniſhing.

MINUTE THE TWENTYFIFTH.

A NARROW VISTA, if the outlines are tolerable, ſhould be free. A ſingle tree, be it ever ſo beautiful, left ſtanding in ſuch a viſta, has a bad effect; as frittering down, and dividing, that which ought to be a whole.

NOR can the eye bear a SINGLE TREE, of an ordinary form, near a well outlined maſs of wood.

THESE are truths which experience taught at this place. The views, in both caſes, were improved, by removing the ſingle trees *.

MINUTE

* Neverthelefs, when the outline of a group, or of a large maſs of wood, is ragged and unſightly, a well featured Tree, ſtanding near its margin, may, by engaging the eye, be advantageous. *See* MIN. 31.

MINUTE THE TWENTYSIXTH.

BUT the moſt intereſting idea, we find regiſtered, in attempting to improve the appearance of this place, aroſe in freeing a grove of full grown Oaks from the foulneſſes with which it was obſcured, and rendered altogether unintelligible, from the principal point of view. Having freed its ſurface, ſufficiently to give a general idea of its figure and outline, and having diſcloſed irregular plots of greenſward on different ſides of it, the deſired effect was produced! though the greater part of the baſe, on the ſide toward the eye, remained foul. By the help of the plots of ground which were ſeen, and by the form of the canopy being obvious to the eye, the imagination readily conceived the reſt, and rendered the whole intelligible.

GENERAL REMARKS.

FROM the ſum of the experience gained, here, we have learnt, that the principal requi ites, in developing the beauties and bruſhing away the deformities of a place overloaded with wood, are circumſpection and application. It is true, an eye habituated

habituated to picturable ſcenery, and eſpecially one which has been accuſtomed to produce it, will diſcover beauties in the midſt of deformities, more readily, than one which is inexperienced; yet, in complicated caſes, the keeneſt cannot decide at ſight.

THIS may account, in ſome meaſure at leaſt, for the "monotony and baldneſs," complained of in MR. BROWN's manner. Towards the latter part of his practice, Mr. B. had but little time to beſtow on any one of the numerous places he was engaged in: if obvious beauties ſtruck him at ſight, in the fortuitous ſcenery of the place to be improved, he no doubt retained them; but he had not time *to ſearch for beauties among deformities*; nor, if he had detected them, had he an opportunity to attend to their developement: he could not be in every place, at the time of foliation or diſcolouring of the leaves; nor had he leiſure to ſtand, in perſon, to watch the effect of the fall of each tree, and thereby determine the fate of the next: and, being thus unable to unravel the knot, he might ſometimes cut it: While at STOWE, BLENHEIM, and FISHERWICK, where he had leiſure to attend perſonally to the minutiæ, he produced a richneſs of ſcenery, which ſhews that he had great abilities in his profeſſion.

SECTION

SECTION THE THIRD.

MINUTES IN PERTHSHIRE.

THE laſt place, to which the ſame purſuit led was TAYMOUTH, the principal reſidence of the EARL OF BREADALBANE, in PERTHSHIRE.

DESCRIPTION OF THE SITE.

TAYMOUTH is ſituated in the center of the Southern Highlands, among the loftieſt and moſt ſavage of the Grampian mountains; but in a fertile and ſpacious valley,—the head of STRATH TAY,—the garden of the Highlands.

THE upper part of the valley, above the grounds of Taymouth, is occupied by LOCH TAY, and its fertile banks; overlooked, on one ſide, by the HILL OF LAWERS, and, at the further extremity is ſeen, in a ſtriking point of view, BEN-MORE; mountains which rank among the very firſt in the iſland.

THE baſe of the valley is about a mile in width, and the culturable lands, which hang on the lower margins of its mountain ſides, about a mile more. Above theſe, lines of green mountain paſture ſucceed; but their greenneſs is preſently loſt, in the heath of the mountains on mountains, which riſe on either ſide of the valley; eſpecially on the North.

THESE MOUNTAINS, though often ſteep, are ſeldom broken or rocky;—except the heads of the Farragan, Schehallion, and Glenlyon hills which are in the higheſt ſtyle of mountain ſcenery. But they are hid, in a great meaſure, from the grounds of Taymouth, by a beautiful hillock, or minor mountain, DRUMMOND HILL, which riſes ſteeply on every ſide out of the baſe of the valley (here dilated), and ſkreens them on the North.

THE RIVER TAY, which receives its name from the Lake, and is, at its efflux, a river of the third or fourth magnitude of rivers in this iſland, takes a winding courſe through the baſe of the valley, which has evidently been formed by water: for although its ſurface, at preſent, is uneven; lying in level ſtages, one higher than another, according to the period of time at which each has been formed; the higheſt ſtages being not leſs than thirty or forty feet above the preſent bed of the river; yet each

each ſtage is of the ſame gravelly loam, which conſtitutes the ſoils of all the "haughs," or river-formed lands of the Highlands.

BUT the time and manner, in which the higher ſtages have been formed, ſeems difficult to be accounted for; unleſs it were done primevally, when the mountains and vallies themſelves were formed; or, during ſome extraordinary convulſion, ſince that period. In Glen-lyon, a narrower and longer valley, there are ſtages of land, evidently water-formed, which cannot be leſs than forty or fifty feet above the preſent bed of the Lyon *.

THE loweſt ſtages of the lands of Taymouth riſe only a few feet above the water, in the time of floods; which, being regulated by the extenſive ſurface of the lake, do not riſe to a great height.

THE HOUSE is ſituated on the largeſt of theſe lower levels, which is nearly encompaſſed by the river; the buildings ſtanding in the center of a ſub-peninſula, which, in outline, reſembles the ſection of a bell, and which forms part of a deer park that contains ſome hundred acres.

 FORMERLY,

* Theſe heights, however, are merely eſtimated by the eye, and may not be accurate.

FORMERLY, a ſumptuous garden ſtretched out in ſtraight lines, from the front of the houſe; vying with Moor Park in ſymmetry and ſweetneſs;—if one may judge from the deſcription of the one *, and a tolerable repreſentation, which has been preſerved of the other. The lines themſelves have long been eraſed; the park now embracing three ſides of the houſe;—an ancient Chateau, modernized.

AT a ſhort diſtance from the North front of the houſe, riſes a ſtately avenue of Limes,—the talleſt and fineſt we have ſeen. But the moſt extraordinary circumſtance of this avenue (if it may be ſo termed), is its form or ground plot; which is that of the letter ∩; occupying the *crown* of the peninſula; the ſemicircular part having, it is probable, traced the banks of the river, at the time of forming: a licence this, which, a century ago, muſt have been audaciouſly heretical.

THE late EARL OF BREADALBANE, who poſſeſſed the eſtate, and made Taymouth his principal reſidence during a length of years, made great alterations in the place; and, conſidering the day in which they were done (near half a century

* See SIR WILLIAM TEMPLE's Account of MOOR PARK, in page 216 of this Volume.

century ago, in the early dawn of rational improvements), they remain, among a variety of other improvements which took place on his extensive estates, proofs of his superior abilities.

THE cardinal improvement, which Taymouth has received from the liberal hand of its late possessor, has been effected by PLANTING; by cloathing the naked hills which overlooked his domain, with wood; particularly the STEEPS OF DRUMMOND, which rise with an ascent, almost unassailable by the human foot; yet are now mantled, from the base to near the summit, with woods of the most luxuriant growth: the lower and middle regions with Oaks, Beeches, and other deciduous Woods, interspersed with Larches and Firs; the upper region with the Highland Pine.

THE Southern banks, which rise to the view from the front of the house, take a very different form: they are composed of rich well turned knolls, rising one behind the other, with dips or flattened stages between them; falling back with easy slopes; so as to form a most *beautiful* style of ground: such as is seldom seen in mountainous districts.

THE highest of these well soiled swells had retained its native covering; chiefly Birch; which had been preſerved for the uſe of the tenants of the eſtate: having been the only *timber*, which, on this part of the eſtate, had eſcaped the ravages of licentious tenantry, and contending clans. On the lower ſwells, much planting has been done: more, indeed, than the quality of the ſoil warranted; and more than picturable effect required.

BUT the moſt ſtriking alteration which the late poſſeſſor made, at Taymouth, and which now marks it, perhaps, diſtinctly from every other place, was that of forming terraces along the brinks of the river banks; whether the feet of theſe banks are now waſhed by the river, or whether it has, in proceſs of time, changed its courſe, and formed oppoſite banks; leaving meadowy grounds between them. In one part, the banks now ſtand ſome hundred yards from each other; a field of ſeveral acres occupying the ſpace between them; lying a few feet above the river, and thirty or more below the terraces which wind on either ſide of it.

THE mode of forming theſe terraces was ſimply that of planting lines of trees, chiefly Beeches, nearly parallel with the brink of the bank; winding as the bank winds, unleſs where the bends are abrupt; and there, the trees take a more eaſy flowing line: of courſe, the width of theſe terrace walks, or lines of ſpace between the trees and the bank,

bank, varies: thirty ſeet may, perhaps, be taken as a mean width. Theſe walks have ever been kept in a ſtate of turf, and mown as graſs walks.

THE banks, though moſtly ſteep and rugged, are hung with wood; which, having riſen above the eye, the river, its banks, and the beauties they contained, with the oppoſite hangs of the valley, and the mountain diſtances which riſe behind them, were, of courſe, ſhut out from the view. Theſe terraces were become little more than wide graſs walks, winding under lines of coppice wood, a mile or more in length.

THIS, however, was more ſtrictly the caſe, on the North ſide of the river. On the ſide next the houſe, the river banks are various in height. The walk dips, in ſome places, down to the lower levels, a few feet above the river; from which it was ſeparated, by a thin line of bruſhwood only; and, in two parts, this had been cleared away: ſo that, in theſe parts, the entire ſurface of the river was ſeen.

THESE terrace walks are ſeparated from the farm grounds, on the North ſide of the river, by a half ſunk wall, and a dwarf hedge;—from the park and paddocks, on the South ſide, by a dwarf wall, with ſquare rough ſtone pillars, at equal and ſhort diſtances. On the outſide of theſe pillars, and this dwarf wall, is fixed a ſlight frame of wood, as a ſupport to five or ſix ropes, ſtretched one above another, as a fence againſt the deer. This fence, however, taken all together, is very unſightly, and is inſufficient.

AT the terminations of theſe terraces, ſtand ornamental buildings, after the manner of Stowe; with others ſcattered on the Southern banks of the valley. They are, in general, in a good taſte, and capable of producing as much effect as uſeleſs buildings generally are*. A ſub-cylindrical tower, or obſervatory, raiſed on the higheſt of the Southern ſwells, and partially hid by the wood of Birches, is by far the moſt pleaſing of theſe buildings.

THE undulating ſurface which has been deſcribed, partially wooded, and receiving additional expreſſion from the tower, a fort, and a temple, placed on conſpicuous parts of it, and the whole backed by mountain knolls, form an agreeable view from the front of the houſe. But, unfortunately, the eating room is the only principal room from which it is ſeen. The drawing room and the library are on the Weſt front; from which, the only

* As RETREATS, ſome of theſe buildings have their uſe. Indeed, grounds extenſive as thoſe of Taymouth, and in a climate uncertain as that of the Highlands, require them

only view was a ſquare graſsplot, of a few acres, hedged in by the tall avenue on one ſide, and by another line of tall trees, in front; without any object to entertain the eye, except a ſingle tree, tolerably well featured, ſtanding near the center of this green area.

PRELIMINARY REMARKS.

UNDER theſe given circumſtances, the leading ſteps of improvements were evident. Firſt, to break the ſkreen which ſtood immediately acroſs the view from the Weſt front, in ſuch manner as to diſcloſe to the view from the windows of the principal rooms, the beſt ſcenery which the low ſituation of the houſe, and the circumſtances of the place, were capable of affording. Next, to open the terrace ſkreens, ſo as to give additional feature to the views from the houſe, and, at the ſame time, to diſcloſe the houſe to the walks, in the beſt point of view; as well as to diſplay the beauties of the terraces to each other. But moſt of all, to ſever theſe ſkreens, in ſuch parts, as command picturable compoſition: and this moſt eſpecially where the river, or the lake, forms a middle ground to mountain diſtances.

DURING

DURING a residence of near two months, in the summer of 1792, these improvements mostly occurred, and, in the spring and autumn of 1793, were in some part executed. By reaching the ground, before the foliation of the trees took place, the skreens were most pervious to the eye, and the exact situation of the breaks were, of course, the best seen; and the progress of the foliation was the most favorable time for catching the best outlines, ragged coppice wood is capable of affording.

BUT these ORNAMENTAL IMPROVEMENTS being secondary to more important objects; namely, those of ascertaining the present state of the RURAL ECONOMY of the HIGHLANDS; and of pointing out the means of their improvement, more particularly THE IMPROVEMENT OF THE ESTATE OF BREADALBANE*; few MINUTES were made on them, at the time. Nevertheless, they did not pass entirely without notice.

MINUTE

* Part of the information, collected in this district, has been presented to the BOARD OF AGRICULTURE, as a REPORT concerning the CENTRAL HIGHLANDS. The whole may hereafter appear, together with such MINUTES in RURAL ECONOMY, as were made at TAYMOUTH.

MINUTE THE TWENTYSEVENTH.

1793. May 17. Began to BREAK THE TERRACE SKREENS, about a fortnight ago. The general effect is equal to expectation; but good outlines are not to be had, from the fortuitous wood of the ſkreens; and the abruptneſs and raggedneſs of the cleared banks offend, and prevent the river from being ſeen, with full advantage.

HOWEVER, by ſhelving the outer brink, ſo as to bend the turf, with an eaſy ſwelling ſlope, from the level of the terrace, and giving ſuitable outlines to the ſide ſkreens,—by planting trees and ſhrubs of different heights, to form banks of foliage, like thoſe which are ſeen in the fortuitous maſſes of park ſcenery,—the viſtas will be made to accord, at once, with the middle grounds of the views, and with the terraces, as immediate foregrounds. The eye, whether it be employed in the general compoſition, or in the place of view from which it is ſeen, will be equally gratified; eſpecially, if the more beautiful of the vegetable tribes be made to aſſimilate with the graſſy carpet of the terrace.

MINUTE

MINUTE THE TWENTYEIGHTH.

MAY 19. Much is to be done by application, attentive, but not too intenſe. By ſauntering leiſurely over the ſite of improvement; prying into each receſs, for latent beauties; and penetrating every pervious part, for more diſtant objects,

IN this way, it has been perceived, that the houſe may be rendered a good object from the North terrace, and the bank of the North terrace a beautiful feature from the houſe. That the diſtant mountain of Lawers may be diſcloſed to the drawing room, and that of Benmore, it is hoped, may be ſhewn, in a happy point of view, from the library. What a proſpect for a ſubject to contemplate! A mountain riſing, at near thirty miles diſtance, his own property, and ſituated near the midway of his eſtate; which reaches from his houſe to this mountain, and from this mountain to the Weſtern ſea,—ſome forty or fifty miles ſtill farther diſtant! No other ſubject, perhaps, *can* enjoy ſuch a view. It ought, therefore, to be diſcloſed, even though ſome of the beauties of Taymouth ſhould be ſacrificed to the diſcloſure *.

MINUTE

* On ſtrict examination, however, it was found, that the library ſtands *a few feet* from the line of view. From the terrace, it is ſeen with advantage.

MINUTE THE TWENTYNINTH.

MAY 19. One of the poſſeſſors of this eſtate, fearful that the Tay ſhould wear away the iſthmus, and carry off his caſtle, fetched five hundred of his tenants out of Argyleſhire, to aſſiſt in making a long high pier of ſtones, to guard it. Tradition ſays, that meal being ſcarce, a famine was brought on, by this ill planned work, before the poor fellows could finiſh their taſk. Had one half of the ſtones been lodged, in a flat ſloping pile, againſt the bank of the river, at the bend, the work would have been more effectual, and have been done with one fourth of the labour; beſides preſerving the natural ſweep of the river; inſtead of giving it an unnatural, and, of courſe, an unſightly turn; and, what is ſtill more diſagreeable, forming a noiſome unwholeſome ſwamp, between the pier and the natural banks; and this, in the immediate neighbourhood of the houſe. It would be worth ſome hundred pounds to have it undone. A hint for thoſe who have ſimilar works to execute.

MINUTE

MINUTE THE THIRTIETH.

MAY 25. It is difficult to open a LINE OF VIEW, between two objects (or between an object and a point of view), which, in the outset, cannot be seen from each other: as in bringing the church of Kenmore within the view from the drawing room. A cautious perseverance, alone, can properly effect it. The best assistance, perhaps, is to suppose a middle point, and, beginning at each end, trace a line towards this supposed point; clearing away brushwood and undergrowth. If, on reaching the midway point, the lines happen to take the same direction, the true line is found: if not, the angle, they make with each other, will show on which hand the true point lies. Or if trees only obstruct, and firm ground only intervene, begin at either object, and trace a random line, until the obstructions are permeated, and the required object can be seen: and, by this false line, endeavour to ascertain the true one.

MINUTE THE THIRTYFIRST.

JUNE 2. In breaking through deep skreens of tall grown trees, and where good outlines cannot be had,

had, a handſome ſingle tree, left at a ſmall diſtance from the maſs, engages the eye, and adds to the general effect *.

MINUTE THE THIRTYSECOND.

JUNE 2. The narrower the viſta, the larger and more diſtinct the object. Benmore, viewed through the contracted viſta from the Weſt-front terrace, appears with infinitely more magnificence and ſtrength of character, than when ſeen from the lower terrace; where the view being wide,—the entire valley of the lake,—the mountain appears proportionally ſmall.

THUS, alſo, the tower of the church, ſeen through the viſtas of the terraces, acquires an importance, and a degree of picturable effect, which, beautiful as it is, it does not produce in broad open views.

THIS appears to be a univerſal law, in viſion; and, perhaps, accounts for the extraordinary ſatiſfaction which the eye receives from contracted views; and for the uſe of ſide ſkreens, in landſcape.

MINUTE

* But otherwiſe, if the contour of the maſs be pleaſing, and the ſingle tree ill formed. See page 382.

MINUTE THE THIRTYTHIRD.

JUNE 9. Every picturable compoſition has its proper POINT OF VIEW. If the eye recede too far from the ſide ſkreens, or lateral bounds of the view, it is cramped, and abridged; if it approach too near them, it becomes broad and ſtaring. This is ſingularly evident on the principal terrace. The viſtas are formed of ſuch widths, as to ſhew the views, at preſent, with the beſt effect, from the walk which leads along the inner margin of the terrace. Viewed from the outer margin or brink, the compoſition is deranged, and the raggedneſs of the bank offends, as being diſcordant with the middle grounds and firſt diſtances that are everywhere caught. It has a ſimilar effect, though not cauſed by the ſame law of viſion, as ſtanding too near a picture.

HENCE the expediency of ſtrewing tufts of ſhrubs on the outer margin of the terrace, to force the ſpectator to the proper diſtance.

OTHER uſes of breaking the greenſward of the terrace (in this part of conſiderable width), are thoſe of giving ſuitable outlines to the extremities of of the ſkreens, and of thickening them between the viſtas; the natural ſkreen, in this part, being thin, and permitting a partial light; which would prevent the full effect of the open views. Theſe relieves of ſhrubs will, alſo, give an immediate richneſs and fullneſs to the foregrounds of the ſeveral compoſitions; which, without them, would, for ſome years, until the ſkreens are fully furniſhed, have remained too bald and meagre; the openings being purpoſely made wider than immediate effect requires. Beſide, they will furniſh an opportunity of evincing the genial climature of the Highlands; in which exotics, of almoſt every kind that bear the open air of this iſland, luxuriate. They will, at the ſame time, give immediate beauty to the place of view, in the flowering tribes; whoſe beauties will begin to fade, and may readily be bruſhed away, when the leſs gaudy plants have acquired ſufficient richneſs and elegance of outline, in the luxuriance of their growth, to fill up, with due effect, the conſpicuous parts of the ſcenery, in which they may hereafter appear.

IN DESIGNING theſe RELIEVES OF SHRUBS, it has been a rule, with reſpect to ſize, to keep them within the fulleſt limits, rather than to exceed them. The broken ground may be readily enlarged, but not eaſily contracted. Beſide, as the *ſhrubs* ſpread forward, the *flowers* will want room on the margins,

gins; ſo that there ought to be an annual enlargement, until the full extent be reached; when the flowers ſhould be diſcontinued: and, finally, the ſhrubs themſelves ſhould be removed, or thoſe of the ſkreens be ſuffered to be overgrown, by their more robuſt and taller neighbours: ſingle ſtandards, or groups, being left in the detached tufts, or the whole cleared away, — as circumſtances will not fail to point out.

MINUTE THE THIRTYFOURTH.

SEPTEMBER 1. The ſteep wooded face of Drummond Hill, when looked up to at a ſuitable diſtance from its foot, has a ſtriking effect. But ſuch a view of it does not occur on the kept walks of the terraces; nor has there ever appeared to be any means of producing ſuch a one, without removing part of the circular avenue; until this morning; when a line of walk was ſtruck out, which will be highly advantageous, for viewing the grandeur of this wooded ſteep, and which will afford an agreeable communication between the Weſt-front terrace and the bridge, without profaning this grand remain of antient gardening; whoſe dampneſs and gloom may ſtill be enjoyed, by thoſe who prefer its ſhade to a more cheerful communication.

WHAT

WHAT a length of acquaintance is required, to aſcertain the beſt advantages of a place. The ſureſt guard againſt miſſing them, is, perhaps, not to execute, before the eye and the judgment are fully ſatisfied——until all doubts have vaniſhed.

MINUTE THE THIRTYFIFTH.

SEPTEMBER 8. To give limits to the width of the lawn of the grand terrace, and for the conveniency of the mowers, ran a wavy path along the ſteep face of the bank (here not leſs than thirty feet high), a few feet below the brink; humouring the relieves of ſhrubs which bend over it, as well as the natural or fortuitous varieties of the ground. Its effects from the oppoſite banks are good; and what was unforeſeen, it will, when the bruſhwood below is grown up, add no mean variety to the ſolitary lounger. A ſtill more ruſſet track, at the foot of this ſteep rugge bank, wildly deviating on the margin of the river, would farther add to the variety.

MINUTE THE THIRTYSIXTH.

OCTOBER 31. Not being able to ſee the real plants put into the broken ground of the terraces, the thought of ſetting up FALSE PLANTS, to prove the

the design, and to serve as a guide to the planter, fortunately occurred. The top wood of the trees lately fallen, afforded boughs of different heights and colours; and, in a few days, a full idea of the effect which may be expected from the living plants, some years hence, has been produced.

By means of these false plants, it was found, that some of the segments of broken ground, formed for the purpose of thickening the skreens between the vistas, were not full enough: there was not room to bring down the banks of foliage, with sufficient slope; they were too steep to please the eye, and for the growth of the lower shrubs, in front. A few feet more, in width, have given the required declivity; so as to be able to proportion the fresh plants to those already in their places, without offending the eye, or cramping the marginal shrubs.

Another material advantage gained by this expedient, is, the planter may now proceed without hazard. He may either take an exact measure, or the bough itself, as a gauge to the required plant; and thus select, with certainty, that which is suitable to the given situation.

In setting up these false plants, it was found that, in order to give them a NATURAL EFFECT, it was requisite

requisite to lean their heads considerably outward: and the same principle holds good, in forming tufts, or banks of foliage, with LIVING PLANTS.

FARTHER IMPROVEMENTS SUGGESTED.

THE alterations, mentioned in the foregoing Minutes, are but the minor part of those which TAYMOUTH required. And it will not be altogether uninteresting to those who have seen the place, or may hereafter visit it, to notice, here, some of the many improvements, of which we conceive it to be still further susceptible.

FIRST, The disgusting park fence, which has been mentioned, is required to be removed, and a foss to be sunk, agreeably to the outline already drawn; including an enlargement of the principal terrace, as a place of view, and to receive a conservatory retreat, or morning room, *furnished* with the most fragrant and beautiful exotic plants *.

SECOND, A removal of the whole, or a principal part, of the unsightly River Break, mentioned in Minute Twentynine, is wanted; in order to cover the

* See REVIEW OF THE LANDSCAPE, &c. page 233.

the offensive swamp which it has formed,—with living water; thus uniting it with the present course of the Tay, in a broad River Bend,—with or without an islet.

THIRD, The Lines of Terrace Trees require to be broken, in different parts; especially to let in the more distant mountain scenery to the house; and to be thinned, in other parts, before their tops are injured, by interfering with each other.

FOURTH, The lines of approach, both from the East and the West, may be altered with happy effect. The proposed lines, in the immediate view from the house, have been drawn.

FIFTH, The deer park may be enlarged with advantage,—by admitting the truly parkish passage between the present lodge and the kitchen garden *.

SIXTH, The beautiful sloping grounds, which rise in front of the house, are capable of much improvement. There are many "harsh lines and angular insertions," which require to be softened; and

* The origin of this admirable idea, however, cannot be claimed by the writer of these remarks.

and much picturable beauty to be disclosed: the lower margins of the hanging woods should fall down in loose festoons, at the foot of the slope; and the ragged Birch woods of the higher knolls be thrown into irregular masses, with grassy glades between them, in the forest style. Much of the soil is rich, and the turf, where it is cleared, remarkably fine. In a district where rich soils and fine turf are so sparingly scattered, it were almost criminal to suffer them to remain encumbered with rough coppice wood, now no longer wanted. Utility as well as Ornament requires the well soiled parts to be cleared from their present roughnesses, and the stoney and less reclaimable parts to be filled up closely with wood; thus converting every part to its proper use, and gaining a happy first distance to the mountain offscape, which rises behind it.

SEVENTH, Carriage roads are much wanted along the slopes, on either side of the valley; especially for the accommodation of strangers. The extreme parts of the grounds are too distant, and their access too difficult, to be assailed by the foot, especially of the delicate.

LASTLY, The House, which is not at present equal to the place, or the estate, on which it stands, requires to be enlarged. A principal front, forming a quadrangle with the present house and wings,

 would

would not only command a fine bend of the Tay, in a ſtriking point of view, as well as the beautiful ſcenery laſt deſcribed, but the mountain view, which has been mentioned as moſt deſirable.

WITH theſe alterations, TAYMOUTH, independent of the additional charms of contraſt, ariſing from the romantic ſcenery and ſavage wildneſſes in its neighbourhood, might rank among the firſt places of the iſland. In magnificence of ſituation, and in picturable ſcenery of the ſofter kinds, it is entitled to precedency. The ſublime, the romantic, or the more ſavage features of pictureſqueneſs, muſt not be looked for from the immediate grounds of Taymouth; though the laſt may ſometimes be caught. As a principal ſummer reſidence of a man of fortune (the remoteneſs of its ſituation apart), there are few places equal to it. Places in general are but limited parts of the diſtricts they lie in, or command;—are hemmed in, on one ſide or another, with inſuperable barriers, or with nuiſances that cannot be removed; whereas Taymouth is a diſtrict within itſelf; and every part may be ranged over at pleaſure,—whether for exerciſe merely, or to enjoy the endleſs variety of view, which the different parts of it are capable of diſcloſing.

INDEX.

I N D E X.

Cascade

INDEX.

D.

E.

F.

Fences

INDEX.

Kent,

INDEX.

Obelisk,

INDEX.

Pruning

INDEX.

Shen-

INDEX.

T.

Train-

INDEX.

END OF THE FIRST VOLUME.

Printed in the United States
By Bookmasters